VOLUME NINETY NINE

SEMICONDUCTORS AND SEMIMETALS

Silicon Photonics

SERIES EDITORS

CHENNUPATI JAGADISH

Distinguished Professor
Department of Electronic Materials Engineering
Research School of Physics and Engineering
Australian National University
Canberra, ACT2601, Australia

ZETIAN MI

Professor
Department of Electrical Engineering and Computer Science
University of Michigan
1310 Beal Avenue
Ann Arbor, MI 48109
United States of America

VOLUME NINETY NINE

SEMICONDUCTORS AND SEMIMETALS

Silicon Photonics

Edited by

SEBASTIAN LOURDUDOSS

School of Engineering Sciences,
KTH - Royal Institute of Technology,
Stockholm, Sweden

RAY T. CHEN

Microelectronics Research Center,
The University of Texas,
Austin, United States

CHENNUPATI JAGADISH

Research School of Physics and Engineering,
Australian National University,
Canberra, ACT, Australia

ELSEVIER

ACADEMIC PRESS

An imprint of Elsevier

Academic Press is an imprint of Elsevier
50 Hampshire Street, 5th Floor, Cambridge, MA 02139, United States
525 B Street, Suite 1650, San Diego, CA 92101, United States
The Boulevard, Langford Lane, Kidlington, Oxford OX5 1GB, United Kingdom
125 London Wall, London, EC2Y 5AS, United Kingdom

First edition 2018

ISBN: 978-0-12-815099-3
ISSN: 0080-8784

Publisher: Zoe Kruze
Acquisition Editor: Jason Mitchell
Editorial Project Manager: Shellie Bryant
Production Project Manager: Abdulla Sait
Cover Designer: Alan Studholme

Typeset by SPi Global, India

CONTENTS

CONTRIBUTORS

Tomohiro Amemiya
Institute of Innovative Research (IIR), Tokyo Institute of Technology, Tokyo, Japan. (ch4)

Shigehisa Arai
Institute of Innovative Research (IIR), Tokyo Institute of Technology, Tokyo, Japan. (ch4)

Alexei N. Baranov
IES, Univ. Montpellier, CNRS, Montpellier, France. (ch1)

Mohamed Benyoucef
Technische Physik, Institute of Nanostructure Technologies and Analytics (INA), Center of Interdisciplinary Nanostructure Science and Technology (CINSaT), University of Kassel, Kassel, Germany. (ch2)

John E. Bowers
Department of Electrical and Computer Engineering, University of California, Santa Barbara, CA, United States. (ch6)

Laurent Cerutti
IES, Univ. Montpellier, CNRS, Montpellier, France. (ch1)

Brian Corbett
Tyndall National Institute, University College Cork, Cork, Ireland. (ch3)

Michael L. Davenport
Department of Electrical and Computer Engineering, University of California, Santa Barbara, CA, United States. (ch6)

Dimitris Fitsios
Centre de Nanosciences et de Nanotechnologies, CNRS, Université Paris Saclay, Palaiseau, France. (ch5)

Yuqing Jiao
Photonic Integration Group, Eindhoven University of Technology, Eindhoven, The Netherlands. (ch7)

Tin Komljenovic
Department of Electrical and Computer Engineering, University of California, Santa Barbara, CA, United States. (ch6)

Ruggero Loi
Tyndall National Institute, University College Cork, Cork, Ireland. (ch3)

James O'Callaghan
Tyndall National Institute, University College Cork, Cork, Ireland. (ch3)

Fabrice Raineri
Centre de Nanosciences et de Nanotechnologies, CNRS, Université Paris Saclay, Palaiseau; Université Paris Denis Diderot, Sorbone Paris Cité, Paris, France. (ch5)

Johann Peter Reithmaier
Technische Physik, Institute of Nanostructure Technologies and Analytics (INA), Center of Interdisciplinary Nanostructure Science and Technology (CINSaT), University of Kassel, Kassel, Germany. (ch2)

Jean-Baptiste Rodriguez
IES, Univ. Montpellier, CNRS, Montpellier, France. (ch1)

Gunther Roelkens
Ghent University-Imec, Technologiepark-Zwijnaarde, Ghent, Belgium. (ch3)

Roland Teissier
IES, Univ. Montpellier, CNRS, Montpellier, France. (ch1)

Eric Tournié
IES, Univ. Montpellier, CNRS, Montpellier, France. (ch1)

Minh A. Tran
Department of Electrical and Computer Engineering, University of California, Santa Barbara, CA, United States. (ch6)

Jos J.G.M. van der Tol
Photonic Integration Group, Eindhoven University of Technology, Eindhoven, The Netherlands. (ch7)

Kevin A. Williams
Photonic Integration Group, Eindhoven University of Technology, Eindhoven, The Netherlands. (ch7)

PREFACE

It can safely be stated that electronics dominated the 20th century, whereas photonics begins to dominate the 21st century. The insatiable need for large bandwidth in data and telecom applications, handheld devices, and internet of things, all of which devour huge amount of energy, cannot be satisfied solely by electronics or photonics. It is in this context silicon photonics is considered as an enabling technology for the next-generation high-bandwidth optical communication systems (from intrachip to long distance) by combining relevant building blocks such as waveguides, filters, couplers, modulators, resonators, detectors, and lasers on silicon. Several of these building blocks can be realized in silicon in a CMOS fab. Silicon being a poor material for optical gain, light sources (lasers) and amplifiers are normally fabricated with III–V semiconductors. The lasers in telecom wavelengths, 1.3 and 1.55 μm, can be propagated via silicon/silicon dioxide waveguides with low losses in the sub-dB/cm ranges. The large difference in the refractive indices of silicon and silicon dioxide enables to confine light produced by III–V materials in submicron or even nanoscale dimensions with high bending capabilities; thereby smaller footprints and large integration densities are facilitated in silicon photonics.

The relevant roadmaps on silicon photonics (David Thomson et al., 2016, *Roadmap on silicon photonics*, J. Opt. 18, 073003 and *2017 Integrated photonic systems roadmap*, AIM Photonics, March 2018) both clearly point out that the need for high bandwidth, energy efficiency, and low latency (data transfer) will be the driving force for silicon photonics that will enable optical interconnects which will gradually outperform electrical interconnects.

This volume collects the state-of-the-art results on achieving silicon photonic components toward fulfilling the promise of silicon photonics.

In Chapter 1, Epitaxial integration of antimonide-based semiconductor lasers on Si, Eric Tournié et al. demonstrate that III-Sb quantum well lasers can be directly grown on silicon and the lasing emission at 1.5–2.3 μm wavelength range operating under CW conditions at room temperature has been achieved. In addition, InAs/AlSb quantum cascade lasers on silicon emitting near 11 μm operating up to 400 K have also been demonstrated. The ability to achieve both telecom and mid-IR wavelengths opens up the feasibility of achieving silicon photonic components for optical communication and sensing applications.

In Chapter 2, III–V on silicon nanocomposites, J.P. Reithmaier and M. Benyoucef describe a very novel approach of embedding III–V quantum dots in a defect-free planar Si matrix. This method is particularly designed to be compatible with CMOS fabrication since the novel hybrid material containing quantum dots in silicon matrix is fabricated prior to subjecting it to CMOS processes. Thereby the hybrid material will still have optoelectronic properties similar to that of III–V materials and yet will be compatible with CMOS processing.

In Chapter 3, Transfer printing for silicon photonics, B. Corbett et al. demonstrate microtransfer printing technique as a flexible and viable technology for integrating several types of components on silicon. By this method, the authors demonstrate stand-alone lasers on silicon, integrated laser and waveguide on silicon, evanescent laser on Si using a tapered coupling, grating coupling photodiodes on silicon/silicon nitride, FTTH (fiber to the home) transceiver array made of III–V on silicon, and an optical link consisting of light emitting diode and a photodiode are some of the examples.

In Chapter 4, Semiconductor membrane lasers and photodiode on Si, Shigehisa Arai and Tomohiro Amemiya focus on the aspect of achieving ultra-low power consumption in optical interconnects. To this end they realize lateral-current-injection-type membrane distributed feedback (DFB) and distributed reflector lasers. They demonstrate a modulation bandwidth of 20 Gbits/s with the energy cost of less than 100 fJ/bit, which is projected to decrease to 30 fJ/bit if the waveguide losses in the optical link and the electrical resistance can be reduced.

In Chapter 5, Photonic crystal lasers and nanolasers on silicon, Dimitris Fitsios and Fabrice Raineri demonstrate physics and technology of high-performance photonic crystal (PhC) nanolasers on silicon platform. Electrically injected photonic crystal nanolasers on Si/SOI circuitry have been demonstrated and have shown the maturity to be integrated in commercial CMOS-integrated nanophotonics. Of particular interest is the implementation of PhC cavity-based optical memory device with a record footprint of 6.2 μm^2 and an actual repetition rate of 5 Gbits/s. Having a switching times <50 ps with 6.4 fJ/bit, they are achieving low energy consumption, high speed, and dense integration in silicon photonic devices as foreseen in the roadmaps described above.

In Chapter 6, Heterogeneous integration of III–V lasers on Si by bonding, Michael L. Davenport et al. present the results arising out of their pioneering work with bonding III–V lasers on silicon/silicon dioxide, thereby allowing the function of light emission to be integrated with silicon photonic circuits. Their state-of-the-art results include low line-width fully integrated mode-locked lasers, DFB lasers, and widely tunable lasers. This technology is more than adequately demonstrated to be amenable for integration on a wafer-scale level and thereby readily available for high-volume and low-cost manufacturing for next-generation silicon photonics systems.

In Chapter 7, InP photonic integrated circuits on silicon, Jos J.G.M. van der Tol et al. introduce the concepts and developments of a new platform, namely, InP Membrane On Silicon (IMOS), for InP nanophotonic integrated circuits. Passive devices include wavelength demultiplexers, couplers, polarization converters, S-bends, and gratings; active devices include optical amplifiers, lasers, photodetectors, and modulators (with electro-optic polymer). Future development on a process design kit (PDK) that contains technology and/or foundry specific information for generating the building blocks (BB) to be integrated and their compatibility with their connectivity (electrical and optical). Their approach provides a powerful route to nanoscale miniaturization and enhanced circuit-level performance for silicon photonics.

As seen above, silicon photonics is truly a multidisciplinary field. The pioneers and experts of silicon photonics have shared their immense knowledge in physics, materials science, and advanced technology related to this field. It is our wish that the readers will benefit largely from their efforts. Everyone involved in silicon photonics will find in this volume solutions to solve one or more problems that one may face today or ideas to advance in their endeavor. We hope that this volume will be a useful reference book for the scientists and engineers striving for large-volume integrated silicon photonic devices and circuits with high performance, large integration density, low footprints, low cost, low energy consumption, and low latency.

Sebastian Lourdudoss
Royal Institute of Technology, KTH, Stockholm
Ray T. Chen
University of Texas at Austin
Chennupati Jagadish
Australian National University, Canberra

CHAPTER ONE

Epitaxial Integration of Antimonide-Based Semiconductor Lasers on Si

Eric Tournié[1], Jean-Baptiste Rodriguez, Laurent Cerutti, Roland Teissier, Alexei N. Baranov
IES, Univ. Montpellier, CNRS, Montpellier, France
[1]Corresponding author: e-mail address: eric.tournie@umontpellier.fr

Contents

1. ANTIMONIDE-BASED COMPOUND SEMICONDUCTORS

Among the III–V semiconductors the so-called "antimonides" refer to the Sb-rich III–V compounds. They include GaSb, InSb, and AlSb which can all be alloyed with InAs to form ternary, quaternary, or even pentanary alloys closely lattice matched to GaSb or InAs substrates. Fig. 1 shows that the III-Sb multinary materials span a large bandgap range from 0.1 eV up to 1.8 eV while still being nearly lattice matched to GaSb. In addition, the position of the band edges displayed in Fig. 2 reveals that III-Sbs offer the opportunity to form a large variety of band alignments, from type-I, where electron and holes are confined in the same material (e.g.,

Semiconductors and Semimetals, Volume 99
ISSN 0080-8784
https://doi.org/10.1016/bs.semsem.2018.08.002

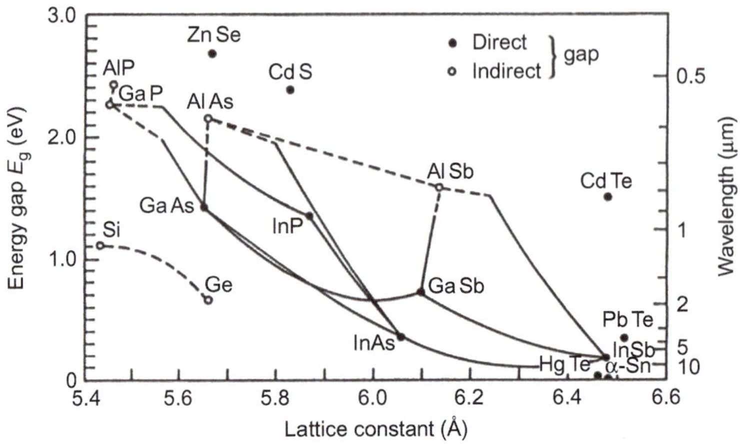

Fig. 1 Energy gap vs lattice parameter of compound semiconductors.

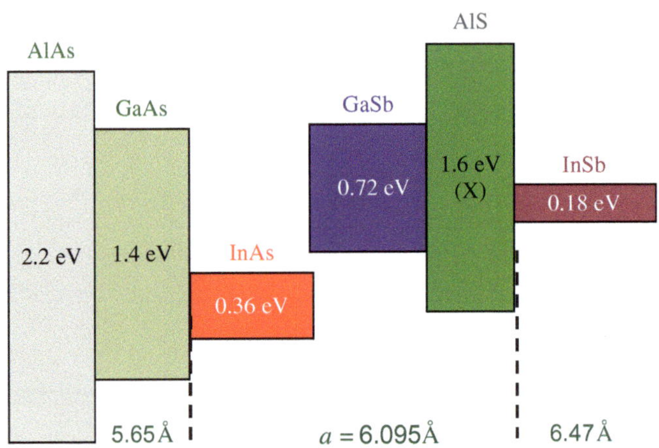

Fig. 2 Band alignment at various III–V interfaces.

AlGa(As)Sb/Ga(In, As)Sb), to type-III, also known as staggered type-II or broken-gap type-II, where the conduction band of one material is located below the valence band of the adjacent one (e.g., GaSb/InAs) through type-II systems (e.g., AlSb/InAs). These properties make III-Sb compounds unique among III–V semiconductors (Vurgaftman et al., 2001). They offer unrivaled opportunities for extensive bandgap and band offset engineering, and for designing devices. In particular, they allow creating artificial, man-made materials whose effective bandgap can be varied by design in the whole range from the near-infrared (IR) to the long-IR.

In the last decade, the III-Sb technology has been extensively studied in view of demonstrating mid-IR optoelectronic devices such as AlGaAsSb/GaInAsSb laser diodes (LDs) for the 1.5–3.3 μm wavelength range (Belenky et al., 2013; Tournié and Baranov, 2012), GaInSb/InAs interband cascade lasers particularly well suited to the 3.5–6 μm range (Vurgaftman et al., 2015), InAs/AlSb quantum cascade lasers (QCLs) covering the whole 3–25 μm wavelength range (Baranov and Teissier, 2015), and InAs/GaSb or InAs/InAsSb high-performance IR photodetectors, progressively challenging the well-established CdHgTe technology both in the mid-IR and longwave IR (Rogalski et al., 2017; Steenbergen et al., 2017). The development of III-Sb laser sources has been driven by numerous applications in the vast field of sensing. Indeed, the mid-IR wavelength range covers several atmospheric transparence windows with fingerprint absorption lines of a number of important chemical species such as alkanes, alkenes, ammonia, BTEX (collective name for benzene, toluene, ethylbenzene, and xylen), VOCs (volatile organic compounds), to name but a few (Rothman et al., 2013). The mid-IR is thus well suited for implementing a variety of photonic sensors that may impact almost every aspect of our society including industrial and environmental monitoring, homeland security, health diagnosis, and many other fields. Until now however all spectroscopic systems are rather bulky.

Photonic integrated circuits (PICs) will provide a route toward low cost and miniaturized spectrophotometer and will therefore be a key technology for mid-IR sensing, provided laser sources can be integrated on Si platforms. This chapter reviews the recent achievements on the way to the epitaxial integration of mid-IR III-Sb-based lasers on Silicon.

2. EPITAXIAL GROWTH OF ANTIMONIDE-BASED COMPOUNDS AND HETEROSTRUCTURES

The epitaxial growth of III-Sb compounds on GaSb or InAs substrates has been intensively investigated since the late 1970s. Most alloys exhibit wide misicibility gaps in very large temperature ranges (Onabe, 1982; Stringfellow, 1982). This is in particular the case for quaternary alloys which are used in most optoelectronics devices.

The growth of III-Sb compounds and devices by metal-organic vapor phase epitaxy (MOVPE) has remained little developed and successful. In fact, the low volatility of Sb, the need of a comparatively low growth temperature, and the strong affinity of AlSb-based compounds with

O and C are difficult issues which render such growth very challenging (Wang, 2004). In addition, there has been less effort toward the MOVPE growth of antimonide compounds than toward other III–V compounds. Recently however, significant progress in this field has been achieved which opens interesting perspectives (Borg et al., 2017; Huang et al., 2017; Wu et al., 2018).

In contrast, much progress has been made in the 1990s in the molecular beam epitaxy (MBE) of compound semiconductors in general, and III-Sbs in particular. It is indeed noticeable that until now all high-performance antimonide-based laser devices have been grown by MBE. An important step forward has been the development of valved cracker cells for group-V elements. The cracker part allows breaking the As_4 and Sb_4 tetramers into As_2 and Sb_2 dimers, respectively, while the needle valve provides a good control of the group-V flux which is particularly important when growing mixed group-V alloys. In addition, the incorporation coefficient of dimers is close to unity while that of tetramers is lower than 0.5 (Foxon and Joyce, 1977). The use of such cells on a routine basis thus results in lower background pressures and in better controlled alloys with a higher crystal quality (Rouillard et al., 1995). Indeed, the control of the group-V compositions in mixed group-V alloys is always a critical issue. This is particularly true with the AlGaAsSb compound which is generally used as several-μm-thick cladding layers in LDs. This imposes a stringent control of the As composition. An efficient way to achieve this is to set the group-III and Sb fluxes needed to grow a pure AlGaSb alloy and then to open the As-valve so as to incorporate the right amount of As to reach lattice matching. Note that this procedure evidently relies on perfectly stable and reproducible group-V valved cracker cells. As for the cladding and contact layers, Te evaporated from Sb_2Te_3 or GaTe sources is used as the n-type dopant while Be is generally used as p-type dopant. The popular amphoteric dopant, Si, dopes most Sb-based semiconductors to be p-type.

Typical MBE growth conditions on native GaSb or InAs substrates are as follows. The Sb/group-III flux ratio should be kept around 2. As with other materials systems, the higher the Al (resp. In) fraction in the alloy, the higher (resp. lower) the growth temperature should be. AlGaAsSb quaternary alloys can be grown at around 520°C while most GaInAsSb alloys are typically grown at between 420 and 470°C depending on the exact composition. GaSb, InAs, and InSb are grown at around 500, 470, and 400°C, respectively.

3. MBE OF ANTIMONIDES ON SI SUBSTRATES

3.1 Introduction

The epitaxial integration of antimonides on highly mismatched substrates has been investigated in view of developing metamorphic optoelectronic devices. In fact, at high lattice mismatches, III-Sbs exhibit peculiar strain relaxation properties, as compared with other III–V compounds. It has been shown that during MBE growth of GaSb on GaAs (lattice mismatch ~8%), strain relaxation can occur by formation of pure edge-type misfit dislocations arranged in a two-dimensional network confined near the III-Sb/substrate interface instead of a high density of 60 degree threading dislocations (Huang et al., 2006; Richardson et al., 2011; Rocher and Snoeck, 1999). This arises from the fact that III-Sb-based materials are relatively soft (Majtykaa et al., 2016), which lowers the energetic barrier to the nucleation of misfit dislocations as compared to stiffer materials such as arsenide, phosphide, or nitride semiconductors. These sessile misfit dislocations, in turn, are the most efficient defects to relieve the strain in highly mismatched materials systems, as previously demonstrated in the InAs/GaAs material system (Trampert et al., 1995). This particular relaxation mode allows fast and efficient strain relaxation, and thus avoids the need for complex and thick buffer layers to reach full relaxation. However, one should bear in mind that it does not preclude the existence of threading defects, in contrast to what is often assumed or even claimed.

A similar behavior has been observed in the MBE growth of GaSb on Si, where the lattice mismatch is as high as 12% and the critical thickness for strain relaxation is below 1 ML (Akahane et al., 2004; Huang et al., 2008; Kim et al., 2006). Fig. 3 shows transmission electron microscopy (TEM) images of III-Sb/Si interfaces which clearly evidence the presence of a regular network of misfit dislocations at the interface (Fig. 3A), and the miscut angle used for the substrate (Fig. 3B). More details on the substrate preparation and nucleation steps are given in Sections 3.2 and 3.3.

3.2 Silicon Substrate Preparation

Growing high quality III–V epitaxial layers on Si is known to be challenging because of the large lattice- and thermal-expansion mismatches and of

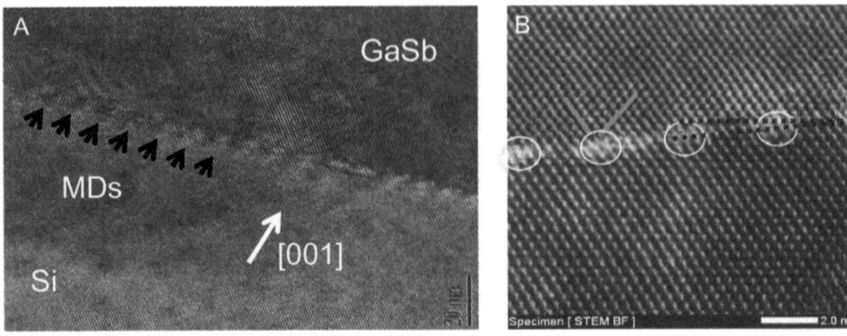

Fig. 3 Cross-section transmission electron microscopy (TEM) images of GaSb/Si interface revealing (A) the presence of a network of 90 degree misfit dislocations at the interface and (B) the miscut angle. *TEM images: Courtesy A. Trampert, PDI-Berlin.*

the polarity difference. This generally results in highly defective layers due to a high density of dislocations, APDs, or twins (Choi et al., 1988; Kroemer, 1987).

Antiphase domains (APDs) form when growing III–V materials on Si due to the polar/nonpolar nature of the III–V/Si interface. In order to obtain APD free III–V layers, either a perfect doubling of all Si surface steps or self-annihilation of all APDs must be achieved. Factors influencing step formation on silicon have been thoroughly studied. In MOVPE reactors, annealing the (001) Si substrate at high temperature under a proper H_2 flow promotes the formation of double step surfaces (Volz et al., 2011). APD free III–V material can then be obtained on on-axis substrates (Cerba et al., 2018; Li and Lau, 2017; Martin et al., 2016). Such conditions however have not been met in MBE systems, yet. On the another hand, double step becomes increasingly stable when rising the substrate disorientation with respect to the nominal (001) orientation (Baski et al., 1997). The established method in MBE growth to promote double step formation, and thereby to suppress APD creation, is to use miscut substrates with an angle larger than around 4 degree toward the [110] direction. All devices reported in this chapter have thus been grown on misoriented substrates.

A crucial point prior to any epitaxy is obviously the substrate preparation. In the case of III–V-on-Si epitaxy, however, this is even a critical issue since the silicon surface is very reactive and metallic and organic contaminants coming from exposure to air, storage boxes, and polishing are easily present on its surface (Habuka and Otsuka, 2001). The MBE community has thus developed various ex situ preparation strategies (Madiomanana et al., 2015). The number of published approaches however demonstrates that this issue

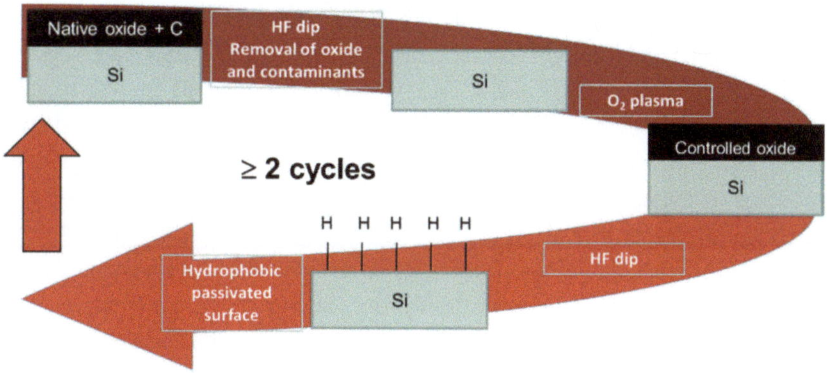

Fig. 4 Ex situ surface preparation of (001) Si substrate prior to III–V epitaxy.

is all but trivial. The important point is that the preparation should remove the native oxide and remove any contaminant. This is done by trapping nonvolatile contaminants in an oxide formed in a controlled manner at the Si surface, this controlled oxide being removed at a later stage by a HF bath which also passivates the Si surface with Si—H bonds. A simple and reproducible Si surface preparation, schematically depicted in Fig. 4, is based on cycles of controlled oxidation/deoxidation which result in a (001) Si surface suitable to III–V epitaxy. However, this does not allow the formation of double steps on the (001) Si surface (Madiomanana et al., 2015). Work remains to be done to solve this issue.

3.3 Nucleation Layer and GaSb Buffer Layer

A perfect surface cleanliness and flatness is thus an important prerequisite to achieve a good epitaxy of the III–V materials on silicon substrates. The next crucial step concerns the very beginning of the growth, and how the III–V material nucleates on the silicon surface. The large chemical energy and lattice mismatch at the interface of the two materials often translate in a three-dimensional growth, together with a strain relieving process occurring in the very first moment of the growth. The material growth then occurs by coalescence of islands and the transition to a two-dimensional growth mode. The two steps, nucleation and 2D growth, are usually optimized separately as they involve different surface or interface energies or strain state. For example, the growth of GaAs on silicon generally involves a radical change in substrate temperature aimed at achieving this nucleation/2D growth sequence (Bolkhovityanov and Pchelyakov, 2008). In the case of GaSb, it was found that the use of an AlSb initiation layer greatly enhances the material quality (Akahane et al., 2004). Fig. 5 shows atomic force microscopy (AFM) images taken from two identical GaSb layers grown on silicon (6 degree-off) with and without the use of an AlSb initial layer. In the first case, an almost poly-crystalline growth is observed, with a measured RMS roughness as high as ~28 nm. In sharp contrast, the predeposition of even a single monolayer thick AlSb layer drastically improves the morphology of the surface, with an RMS roughness down to 4.7 nm. AlSb creates faceted islands on the silicon surface which serve as nucleation sites for the growth of GaSb, and by decreasing the Ga diffusion length the islands also facilitate the transition toward a bidimensional GaSb layer. Fig. 6A and B shows an example of such an AlSb island grown on a 6 degree-off Si substrate and covered by GaSb imaged by TEM-EDX. The corresponding Ga- and Al-contrasts are clearly visible, as well as the different

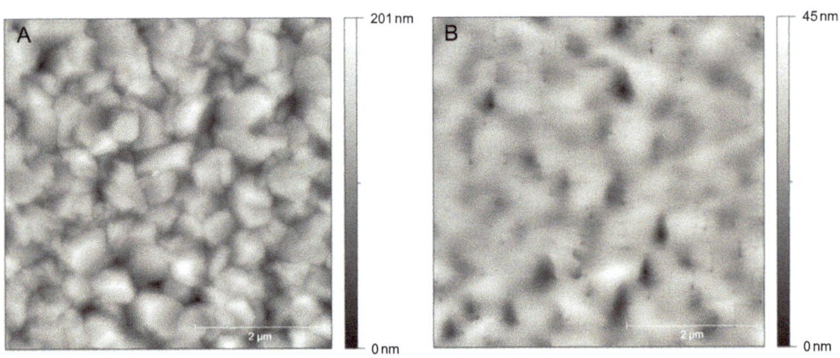

Fig. 5 AFM images taken from a GaSb layer (A) directly grown on Si and (B) grown on an AlSb nucleation layer.

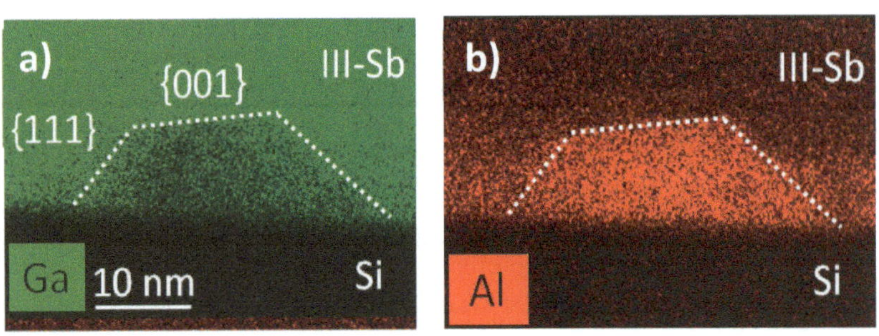

Fig. 6 TEM-EDX images of an AlSb island grown on a 6 degree-off Si substrate covered by GaSb illustrating (A) Ga contrast and (B) Al contrast. *TEM: Courtesy G. Patriarche, C2N, CNRS.*

facets surrounding the island. High-resolution TEM image of the interface between Si and AlSb also reveals a highly ordered interfacial misfit dislocation network (Fig. 3).

The morphology and density of the AlSb islands depend on several growth parameters, among which the total amount of AlSb deposited and the substrate temperature plays a key role. The influence of these two parameters on the full-width at half-maximum (FWHM) of the GaSb peak measured by high-resolution X-ray diffraction (HR-XRD) is shown in Fig. 7 (Rodriguez et al., 2016). The data were taken on omega-scans measured on identical structures grown on 6 degree off-axis silicon substrates and comprising an AlSb nucleation layer, a 500-nm-thick GaSb buffer layer and a 500-nm-thick quantum-well (QW) structure based on GaInAsSb/AlGaAsSb layers. The three curves have similar shapes, and reveal an optimum AlSb thickness evolving from ~1–4 MLs at 400°C to ~17 MLs at 500°C. We propose that for each substrate temperature, this value corresponds to the AlSb amount required to reach the maximum island density at the silicon surface. In this scenario, a maximum density of nucleation site for the growth of GaSb is reached, and the transition to a 2D growth is rapidly achieved through the change of material deposited, namely from

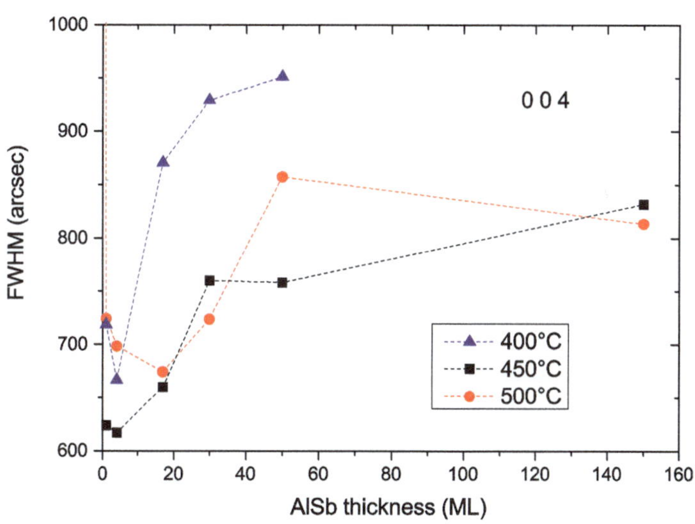

Fig. 7 Variation of the FWHM of the 004 GaSb peaks with the AlSb nominal thickness for different substrate temperature during the nucleation layer growth. *Reprinted from Rodriguez, J.-B., Madiomanana, K., Cerutti, L., Castellano, A., Tournié, E. 2016. X-ray diffraction study of GaSb grown by molecular beam epitaxy on silicon substrates. J. Cryst. Growth 439, 33–39, with permission from Elsevier.*

AlSb to GaSb. It is worth mentioning that since the pioneering work of Akahane et al. (2004) most people, including ourselves, used to grow a 17 ML AlSb nucleation layer at 500°C. Even though this corresponds to a local optimum, this is not the *optimum optimorum* (Fig. 7).

Annealing of highly mismatched heteroepitaxial structures has been known for a long time to be an efficient way of improving the material quality (Ayers et al., 1992; Yamaguchi et al., 1990). Reduction of the dislocation density arises due to their annihilation caused by the dislocation movement and coalescence at high temperature. This effect of annealing was studied on the GaSb/AlSb on 6 degree off-axis silicon substrates heteroepitaxial structures using a set of samples with a 5-nm-thick AlSb nucleation layer and GaSb layers with thicknesses of 0.1, 0.2, 0.5, and 1 μm, all grown at 500°C. X-ray rocking curves have been measured on as-grown samples, which were then reloaded in the MBE reactor for being annealed at 550°C under Sb flux during 30 min to 1 h, depending on the layer thickness. A comparison of the results before and after complete annealing is displayed in Fig. 8. The improvement brought by annealing increases with the GaSb layer initial thickness, from a reduction of about 7% of the FWHM for the thinnest sample to about 24% for the 1-μm-thick layer. Therefore, while

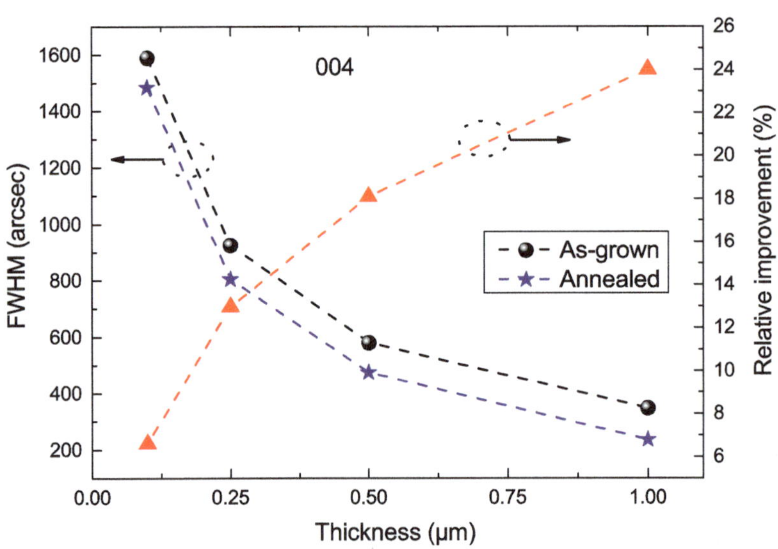

Fig. 8 Comparison of the FWHM improvement versus the GaSb thickness after 2 h annealing at 550°C. *Reprinted from Rodriguez, J.-B., Madiomanana, K., Cerutti, L., Castellano, A., Tournié, E. 2016. X-ray diffraction study of GaSb grown by molecular beam epitaxy on silicon substrates. J. Cryst. Growth 439, 33–39, with permission from Elsevier.*

an annealing step at the initial stage of the buffer growth seems to only have a marginal effect on the material quality, the improvement becomes quite significant for thicker buffer layers. After a complete annealing process, the omega-scan of a 1 μm GaSb layer exhibits an excellent FWHM of 235 arcsec, down from the 347 arcsec measured on the as-grown sample (Rodriguez et al., 2016). Such FWHM compares favorably with values obtained for the same thickness in the case of the epitaxy of Ge on silicon (~160–200 arcsec, Shin et al., 2010) and GaAs on silicon (~200 arcsec, Bolkhovityanov and Pchelyakov, 2008).

Such annealed layer mimics the buffer layer of a laser structure which gets annealed during the whole laser growth.

4. GaInAsSb/AlGaAsSb LDs GROWN ON Si

As mentioned in Section 3.1, GaSb-based LDs generally rely on GaIn(As)Sb/AlGa(As)Sb QWs which have demonstrated high performances between 2 and 3.3 μm when grown on GaSb substrates (Belenky et al., 2013; Tournié and Baranov, 2012).

When grown on Si substrates, optically pumped LDs have been demonstrated with AlGaSb/GaSb/AlGaSb double heterostructures as early as 1986 (van der Ziel et al., 1986), but the first electrically pumped LDs were reported 10 years later (Jallipalli et al., 2007). The structure was based on 10-nm-wide GaSb QWs confined by $Al_{0.3}Ga_{0.7}Sb$ barrier layers. Growth was performed at low temperature (400°C) on 5 degree-off (001) Si substrates to reduce the formation of APDs. Prior to the growth, the surface of Si substrates was simply hydrogen passivated by immersing the wafer in a buffered HF bath. A thermal cycle at 800°C was applied prior to growth initiation to ensure removal of H. A 50-nm-thick AlSb nucleation layer was inserted between the Si substrate and the laser structure. Lasing was achieved under pulsed conditions at 77 K with a threshold current density near 2 kA/cm^2 and a wavelength of 1.55 μm (Jallipalli et al., 2007).

The next objective was to reach continuous wave (cw) operation above room temperature (RT). We demonstrated RT operation with a laser structure grown on a 6 degree off n-type Si substrate (Rodriguez et al., 2009). The buffer layer was grown at 510°C and started by a 5 nm AlSb nucleation layer, as described earlier (cf. Section 3.3), followed by a 1-μm-thick GaSb:Te layer. The remaining part of the laser structure was similar to that used on GaSb substrates. The active region was composed of two compressively strained 11-nm-wide $Ga_{0.65}In_{0.35}As_{0.06}Sb_{0.94}$ QWs confined by $Al_{0.35}Ga_{0.65}As_{0.03}Sb_{0.97}$ barrier layers and it was embedded in between 1.5-μm-thick $Al_{0.9}Ga_{0.1}As_{0.0.7}Sb_{0.93}$ n- and p-type cladding layers. 300-nm-thick spacers with the same composition as the barrier layers were inserted between the claddings and the active region to form the waveguide. A p^{+}-GaSb contact layer completed the structure. The bandgap offsets were smoothed out by inserting graded composition layers between the claddings and the top and bottom GaSb layers. Au–Ge–Ni was used as the n-type contact metal on the Si substrate backside. Fig. 9 shows the typical light–current–voltage (L–I–V) characteristics measured on a LD under pulsed operation at various duty cycle (1% and 5%) at RT. The threshold current density was measured to be around 1.5 kA/cm^2, a factor ~15 higher than for similar lasers grown on native GaSb substrates (Salhi et al., 2004). This high value could be explained by high internal optical losses related to residual threading dislocations. An output peak power in the range of a few tens milliWatt was measured. The voltage characteristics presented a turn-on voltage of 2.8 V, much larger than the 0.7 V measured on the typical lasers grown directly on GaSb. This was attributed to the presence of a high defect density at III-Sb/Si interface which degrades dramatically the electrical performances. Moreover, Fig. 9 evidences that the threshold current density increases while both the serial resistance and the external quantum efficiency decrease when the duty cycle increases, indicating adverse thermal properties. The inset in Fig. 9 shows the laser spectrum with the main peak emission at 2.33 μm.

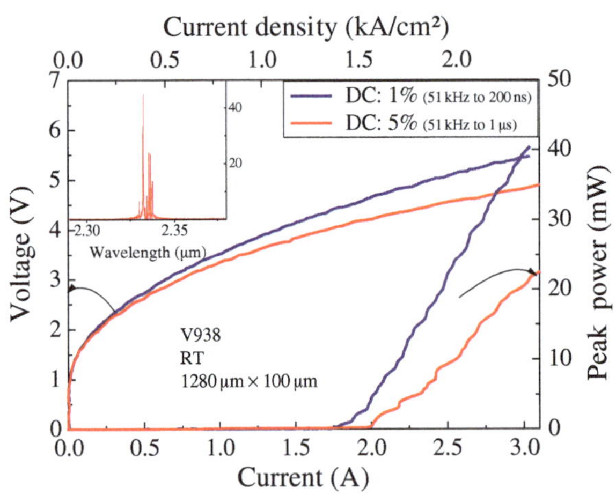

Fig. 9 Room temperature P–I–V characteristics at 2.3 μm for different duty cycle. The *inset* shows the lasing spectrum of the device at RT.

The LD presented earlier had the current driven through the highly defective Si/III-Sb interface which resulted in degraded performances. To overcome this issue, a new process was derived where an $InAs_{0.92}Sb_{0.08}$ layer was inserted within the GaSb buffer layer. $InAs_{0.92}Sb_{0.08}$ is lattice matched to GaSb while n-type layers exhibit a very low contact resistance and a high electrical conductivity as compared to GaSb (Lauer et al., 2006). In addition, it offers a perfect selectivity with respect to GaSb for wet etching (Dier et al., 2004). The new buffer layer sequence was then: 5 nm AlSb/ 200 nm GaSb/150 nm InAsSb/800 nm GaSb. After completion of this n-type, Te-doped, composite buffer layer a typical laser structure designed to emit at 2 μm was grown, with 1.5-μm-thick $Al_{0.9}Ga_{0.1}As_{0.07}Sb_{0.93}$ cladding layers and 200-nm-thick $Al_{0.25}Ga_{0.75}As_{0.02}Sb_{0.98}$ waveguide layers. The active region was made of two compressively strained 9-nm-wide $Ga_{0.65}In_{0.35}As_{0.05}Sb_{0.95}$ QWs separated by 30 nm of the same material than the waveguide. In this configuration, the overlap between the narrow gap InAsSb and the fundamental optical mode was very weak, resulting in negligible optical losses (Reboul et al., 2011). Ridge LDs were then processed using standard photolithography and wet etching, with the n-contact taken on the InAsSb layer located within the buffer layer. Threshold current densities as low as 850 A/cm^2 in pulsed mode at RT were obtained with 100 μm × 1.4 mm Fabry–Perot cavities, to be compared with the 1.5 kA/cm^2 in the standard contact scheme. CW operation was demonstrated with 8 μm × 2 mm narrow ridge cavities. Fig. 10 presents the L–I–V curves taken in cw mode at various temperatures. The turn-on voltage was now close to 0.8 V, a value comparable to that of LDs grown on GaSb substrates and emitting at the same wavelength (Garbuzov et al., 1996; Salhi et al., 2004). This demonstrated that driving the current through the III-Sb/Si interface was indeed a limiting factor in the conventional contact scheme. The cw output power measured at 350 mA varied between 8 and 2 mW/facet when the temperature was ramped from 10 to 35°C. The external quantum efficiency (η_d), deduced from the slope of the L–I curves, changed from 16% to 12% in this temperature range. Note that η_d remained constant for each measured temperature, which showed that the thermal rollover had not been reached. The characteristic temperature T_0 that characterizes the evolution of the threshold current intensity with the temperature, varied from 80 K below 20°C to 40 K at higher temperatures, values slightly lower than that obtained with similar structures grown on GaSb substrates (Garbuzov et al., 1996; Salhi et al., 2004). In addition, the still high threshold current density (850 A/cm^2 vs 100 A/cm^2 for LDs grown on GaSb substrates) was

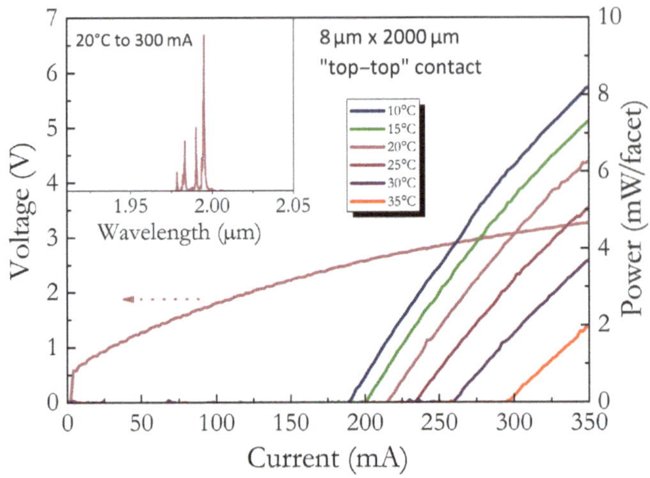

Fig. 10 P–I–V characteristics for various temperature of a narrow ridge 2 μm laser diode grown on Si and with n- and p-type contact on the epitaxial side in CW regime. The *inset* shows CW lasing spectrum at 20°C with a driving current of 300 mA. *Reprinted from Reboul, J.R., Cerutti, L., Rodriguez, J.B., Grech, P., Tournié, E. 2011. Continuous-wave operation above room temperature of GaSb-based laser diodes grown on Si. Appl. Phys. Lett. 99, 121113, with permission from IP.*

attributed to the high internal losses. Indeed, a value of 20 cm^{-1} was determined from Hakki–Paoli measurements (Reboul et al., 2011), around five times higher than the best results obtained for lasers grown on GaSb. As mentioned above for previous III-Sb lasers grown on Si, these high optical losses are mainly attributed to optical scattering within the waveguide due to the residual threading dislocations originating from the III-Sb/Si interface. Finally, the inset in Fig. 10 displays the main mode peaks at 2 μm under 300 mA cw current drive at 20°C.

These results thus show that GaSb-based LDs grown on Si offer the potential for high performances in the traditional wavelength range of this technology. Still, a careful engineering of the QW band structure allows reaching the telecom wavelength range, a core application field of Si photonics. Given the GaSb bandgap (0.725 eV, i.e., 1.65 μm, at RT), the challenges in that case are to design a structure with, on the one hand, wide-enough QWs to preserve a sufficient optical mode/QWs overlap and a sufficient energy level confinement, and, on the other hand, highly strained QWs, i.e., high In contents in the QWs, to favor laser emission (Adams, 2001). CW laser operation at RT could be achieved only recently with such devices based on an active region composed of 3.6-nm-wide

$Ga_{0.8}In_{0.2}Sb$ QWs confined by $Al_{0.35}Ga_{0.65}As_{0.03}Sb_{0.97}$ barrier layers. The cladding, the waveguide, and the contacts layers were identical to that of the 2.3 μm laser described earlier. Grown on a GaSb substrate, this structure showed cw operation up to 45°C with an emission around 1.57 μm, demonstrating that III-Sbs are suited for telecom photonics (Cerutti et al., 2010). The same structure was then grown on a Si substrate and processed into 100 μm × 630 μm LDs. The process relied on an n-type contact taken on the back of the Si substrate. Fig. 11 shows the L–I–V characterization at both 90 K and RT in pulsed regime (100 ns to 21 kHz). The threshold current densities were 0.75 and 5 kA/cm^2, respectively. As discussed earlier, the high turn-on voltage of 3 V can be ascribed to the poor electrical conductivity at the interface between the Si and the III-Sb heterostructure. The inset in Fig. 11 presents the laser spectra taken at 90 K and RT in pulsed mode with drive currents of 0.5 and 3.5 A, respectively. The emission wavelength shifted from 1.42 at 80 K to 1.55 μm at RT, the target wavelength for telecom applications.

Reaching more efficient III-Sb LDs emitting near 1.55 μm required further refining of the active regions and applying optimized processing. The original concept was to insert monolayer thin (0.45 nm) $Al_{0.68}In_{0.32}Sb$ barrier layers within the $Ga_{0.8}In_{0.2}Sb$ QWs. Inserting two such barrier layers within the GaInSb QWs allowed increasing the QWs width from

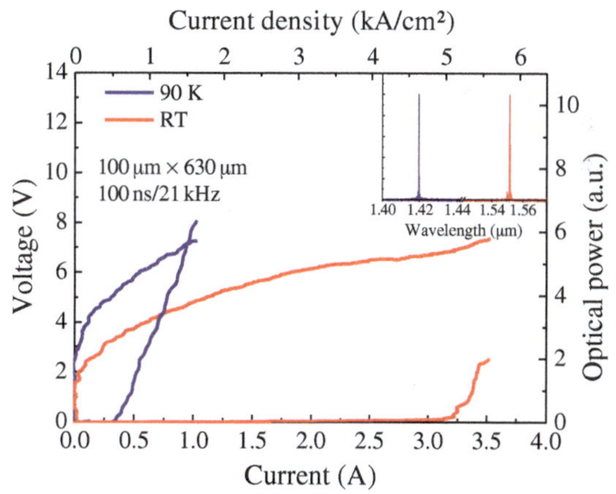

Fig. 11 L–I–V characteristics for a broad area 1.55 μm laser grown on Si substrate at 90 K and RT in pulsed regime. The *inset* shows pulsed lasing spectra under 0.5 and 3.5 A driving current at 90 K and RT, respectively. *Reprinted from Cerutti, L., Rodriguez, J.B., Tournié, E. 2010. GaSb-based laser, monolithically grown on Si, emitting at 1.55 μm at room temperature. IEEE Photon. Technol. Lett. 22, 553–555, with permission from IEEE.*

3.6 to 6.9 nm, the compressive strain from 1.24% to 1.35%, the wavefunction overlap from 94% to 96.2%, and the optical mode overlap with the QWs from 2.6% to 5.2% (Cerutti et al., 2015). LDs with such an active zone grown on GaSb substrates operated in cw above RT with an emission wavelength of 1.55 μm. Moreover, these composite QWs showed a positive impact on the laser properties with a threshold current reduced by a factor of 2 and the characteristic temperature improved from 29 to 72 K in comparison with lasers using only $Ga_{0.8}In_{0.2}Sb$ QWs.

Taking into account the different optimizations described earlier, an optimized III-Sb structure designed for laser emission near 1.55 μm was grown on Si substrate. It was composed of three composite QWs, with two $Al_{0.68}In_{0.32}Sb$ barriers in each $Ga_{0.8}In_{0.2}Sb$ QWs, and with an $InAs_{0.92}Sb_{0.08}$ layer inserted within the GaSb buffer to avoid driving the current through the Si/III-Sb interface. The nucleation layer was 4 ML AlSb grown at 450°C. The structure was then processed using standard lithography and wet etching, the p- and n-contacts being taken in the epitaxial structure. L–I–V curves in pulsed mode at RT for a 100 μm × 1 mm cavity showed current densities around 1 kA/cm^2, five times lower than the previous laser grown on Si substrate and emitting at 1.55 μm. The turn-on voltage was around 0.8 V, close to the bandgap energy and to the value obtained with identical structures grown on GaSb substrates. Processing this structure into 10 μm × 1 mm cavity allowed reaching CW operation above RT. Fig. 12 shows the L–I–V characteristics at various temperatures between 15 and 35°C. The threshold current varied from 300 to 450 mA. The resulting T_0 of 50 K was close to the value obtained for 2 μm III-Sb laser grown on Si substrate and also to the early-generation 1.55 μm InP-based QW lasers grown on InP (Agrawal and Dutta, 1993). The cw output power measured at 500 mA was 3 mW/facet at 20°C and η_d varied from 2.5% to 1% in the 15–35°C temperature range. Moreover, it can be observed that the thermal rollover had not been reached indicating that even higher optical power could be achieved. The laser spectrum displayed in the inset of Fig. 12 shows a peak wavelength at 1.59 μm in cw at 15°C, in the center of the optical communications L-band. This wavelength is longer than that of the same structure grown on GaSb due to the additional tensile strain induced by the different thermal-expansion coefficients of GaSb and Si.

In the last decade, much improvement has been made on GaSb-based LDs grown on Si. In particular, the ratio of the threshold current density measured on broad area LDs gown on GaSb substrate to that of identical LDs grown on Si substrates has decreased from ~15 to ~3, which is

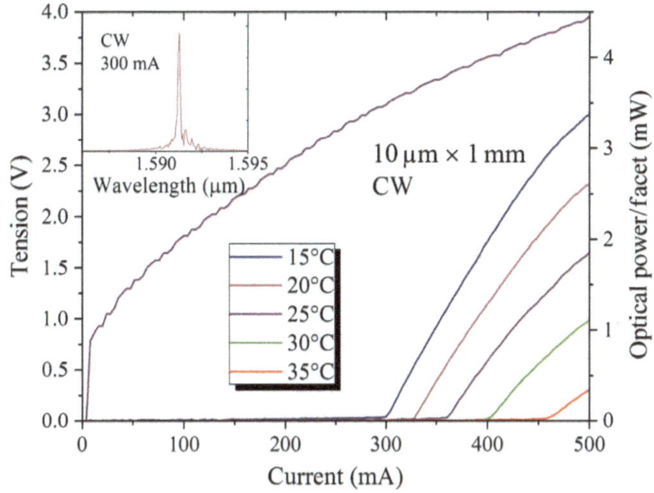

Fig. 12 L–I–V characteristics in cw mode for various temperatures of a narrow ridge 1.55 μm laser diode grown on Si and processed with n- and p-type contacts on the epitaxial side. *Reprinted from Castellano, A., Cerutti, L., Rodriguez, J.B., Narcy, G., Garreau, A., Lelarge, F., Tournié, E. 2017. Room-temperature continuous-wave operation in the telecom wavelength range of GaSb-based lasers monolithically grown on Si. APL Photonics 2, 061301; used in accordance with the Creative Commons Attribution (CC BY) license.*

remarkable. Still, these lower performances are attributed to residual threading dislocations in the structures. Indeed, the density measured is in the range of $10^8\,cm^{-2}$ which is probably too high to reach long life time laser operation (Jung et al., 2018). The next challenge is to reduce the defect density, e.g., via dislocation filtering strategies, to open the way to long-lived laser sources.

5. InAs/AlSb QCLs GROWN ON Si

QCLs exhibit a number of advantages making this technology extremely attractive for developing integrated MIR sensing systems: it covers an extremely large spectral range from ~3 μm up to the THz domain (Razeghi et al., 2015; Yao et al., 2012), this is the most energetically efficient laser technology (Bai et al., 2010; Liu et al., 2010), and it supports frequency combs (Rösch et al., 2015). Integrating QCLs on Si is thus a crucial challenge on the way to smart sensing systems.

The InAs/AlSb material family is very attractive for use in QCLs due to the high conduction band offset and the small electron effective mass favorable to obtain high intersubband optical gain. Lasers with record performances based on these materials were demonstrated, such as QCLs operating in the cw regime at RT above 15 μm—the longest RT cw emission wavelength of semiconductor lasers (Baranov et al., 2016), and pulsed QCLs operating above RT at 20 μm—the longest emission wavelength of semiconductor lasers at RT (Bahriz et al., 2015).

We reported early 2018 the first ever QCL grown on a Si substrate (Nguyen-van et al., 2018). The active zone was based on a design with vertical transitions in four coupled QWs. It consisted of 40 repetitions of the following InAs/AlSb layer sequence: **21**/96/**2.8**/76/**2.9**/73/**3**/70/**6**/64/**7**/62/**7**/58/**9**/<u>57</u>/**14**/<u>56</u>/**17**/55, in Å and starting from the injection barrier, where AlSb layers are in bold and the Si-doped InAs layers ($n = 4 \times 10^{16}\,cm^{-3}$) are underlined. A plasmon-enhanced dielectric waveguide of the laser was formed by 2-μm-thick cladding layers made of n^+-InAs doped with Si to $2 \times 10^{18}\,cm^{-3}$. In order to reduce the overlap of the guided mode with the absorbing doped material and to minimize the propagation losses the active zone with a total thickness of 3 μm was separated from the cladding layers by 2.5-μm-thick undoped InAs spacers. The electromagnetic modeling of the guided modes, using a finite element solver, gives an overlap of the fundamental mode with the active region $\Gamma = 56\%$ and the waveguide loss $\alpha_w = 3\,cm^{-1}$ (not including losses in the active region).

The structure benefited from the nucleation and processing development presented earlier. An InAs-on-Si template was first prepared. The substrate was a (100) Si substrate with a 6 degree miscut toward the [110] direction to limit the formation of antiphase domains appearing during the growth of III–V materials on nonpolar group IV substrates. Prior to epitaxy the Si substrate was prepared by applying both ex situ and in situ procedures described earlier (cf. Section 3.2). The growth was initiated by depositing four monolayers AlSb directly on the Si substrate at 450°C, followed by the growth of a GaSb buffer layer while ramping the substrate temperature up to 500°C (cf. Section 3.3). After 1 μm GaSb, the temperature was ramped down to 450°C in order to grow a 200-nm-thick InAs layer. The QCL growth was then performed on this template using the standard procedure employed usually to grow InAs/AlSb QCLs (Baranov and Teissier, 2015). The same QCL structure was grown side-by-side on an InAs substrate in the multiwafer Riber 412 system.

The wafer was processed into ridge lasers using wet etching and conventional UV photolithography. The ridge width w varied between 10 and 22 μm. Electrical insulation was provided by hard baked photoresist. After processing, the substrate was thinned down to 50 by mechanical polishing.

The laser ridges were etched down to the first n^+-InAs cladding layer where the bottom electrical contact was formed. The second contact was fabricated on the top of the ridges. The contacts to the devices were made using non-alloyed Ti/Au metallization. The contact pads were placed on the opposite side of the ridges. The ridges were oriented along the ⟨011⟩ direction and the atomic plane steps of the epitaxial layers produced by the misorientation of the substrate formed a staircase running perpendicularly to the ridge axis.

The fabricated lasers exhibited RT pulsed threshold current densities *Jth* as low as 1.3 kA/cm^2 and peak optical powers exceeding 100mW/facet for the longest, 3.0 mm long, devices. The threshold current density increased up to 2.4 kA/cm^2 in the shortest, 0.6 mm long lasers. Voltage–current and light–current characteristics of the QCLs are shown in Fig. 13A. The QCL emission wavelength typically increased from 10.5 μm at 80K to 11.1 μm at 380K (Fig. 13B). Fig. 14 presents voltage–current and light–current characteristics of a 3-mm-long laser measured between 80 and 380K. The threshold current density increased slowly with temperature between 80 and 160K whereas above 180K its temperature dependence was exponential with a characteristic temperature $T_0 = 150$K.

It is necessary to note that QCLs grown on Si exhibited similar high performances as the same devices grown on a native InAs substrate. The 3-mm-long QCLs on Si demonstrated *Jth* only 30% higher than that of the lasers grown on InAs and the slope of the light–current curves was also comparable in both types of lasers. Moreover, in the short devices *Jth* was the same or even lower than in the reference QCLs, which indicates a higher optical gain in the lasers grown on Si.

The high performance of the QCLs grown on Si may seem surprising at first sight, especially taking into account the poor crystalline quality of the wafer. Indeed, the surface morphology of the grown wafer was poorer than that of typical QCL structures grown on native InAs substrates and the density of misfit dislocations arising from relaxation of the large lattice mismatch (~11.5%) between the Si substrate and the InAs-based structure was estimated to be (1–3) × 10^7 cm^{-2} (Nguyen-Van et al., 2018). One should bear in mind however that the very short recombination lifetimes involved in QCLs makes them insensitive to many parasitic recombination mechanisms existing in other types of semiconductor lasers (Sirtori and Teissier, 2010). Work is needed in the future to assess these properties in more details.

Given the wide wavelength range reachable with InAs/AlSb QCLs, the demonstration of high-performance lasers grown on Si opens the way to the development of a variety of mid-IR systems.

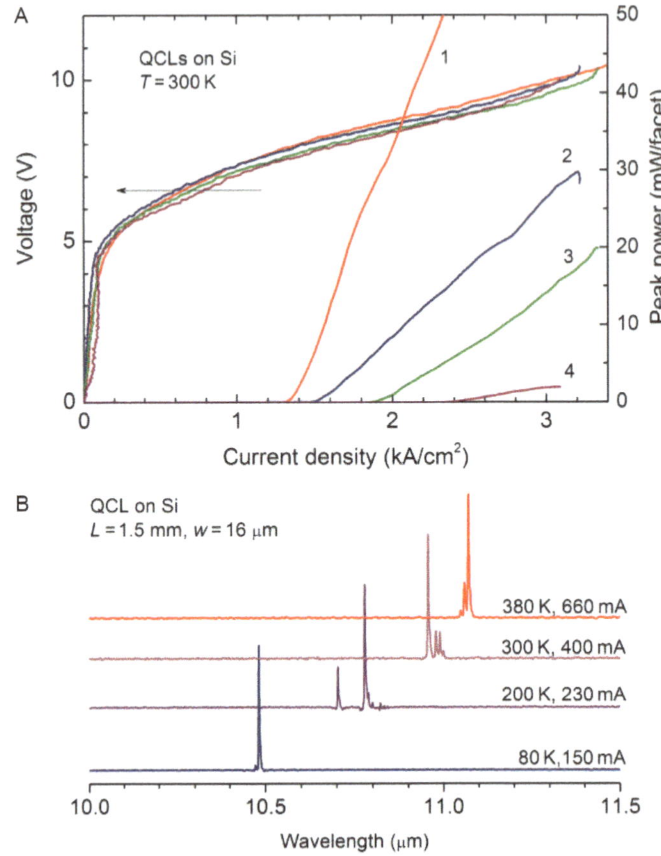

Fig. 13 Characteristics of QCLs grown on Si. *L*: resonator length, *w*: ridge width. (A) Voltage–current and light–current characteristics of various QCLs grown on Si. 1: $L=3.0$ mm, $w=20\,\mu$m; 2: $L=1.5$ mm, $w=16\,\mu$m; 3: $L=1.15$ mm, $w=15\,\mu$m; 4: $L=0.6$ mm, $w=14\,\mu$m. (B) Emission spectra of a QCL grown on Si measured at different temperatures. *Reprinted from Nguyen-Van, H., Baranov, A.N., Loghmari, Z., Cerutti, L., Rodriguez, J.-B., Tournet, J., Narcy, G., Boissier, G., Patriarche, G., Bahriz, M., Tournié, E., Teissier, R. 2018. Quantum cascade lasers grown on silicon. Sci. Rep. 8, 7206; used in accordance with the Creative Commons Attribution (CC BY) license.*

6. OUTLOOK

We have shown in this chapter that the epitaxial integration of antimonide-based semiconductor and devices has made substantial progress in the last decade and room temperature operation has been achieved from 1.5 μm to >10 μm with LDs and QCLs. This makes this technology unrivaled.

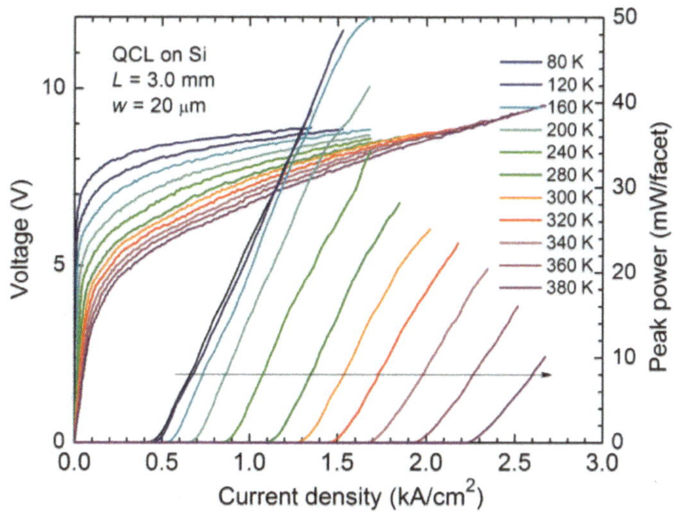

Fig. 14 L–I–V characteristics measured between 80 and 380 K for an InAs/AlSb QCL grown on Si. *Reprinted from Nguyen-Van, H., Baranov, A.N., Loghmari, Z., Cerutti, L., Rodriguez, J.-B., Tournet, J., Narcy, G., Boissier, G., Patriarche, G., Bahriz, M., Tournié, E., Teissier, R. 2018. Quantum cascade lasers grown on silicon. Sci. Rep. 8, 7206; used in accordance with the Creative Commons Attribution (CC BY) license.*

Still, work remains to be done in several directions before getting to practical, long lifetime devices or PICs. Of utmost importance (Jung et al., 2018), threading defect filtering strategies have to be implemented. Next, work should be done toward promoting the MBE growth of APD free buffer layers on on-axis (001)Si since this has been achieved in MOVPE growth only until now. Indeed, for a number of applications devices grown on on-axis substrates may be needed. Note however that this may depend on integration strategies and several options can be envisioned, depending on the target application. Finally, the device design should be adapted to take into account the peculiarity of the III-Sb on Si technology.

Altogether, we are confident that given the unrivaled potential of the III-Sb technology in the mid-IR, our results will open the way to the development of a large portfolio of integrated sensors.

ACKNOWLEDGMENTS

Part of the work performed in our group and referred to in this review has been supported by the French program on "Investments for the Future" (EquipEx EXTRA, ANR-11-EQPX-0016), the French ANR, *Régions Languedoc-Roussillon* and *Occitanie*, and the European Union. We should like to thank our colleagues whose names appear in the references for their collaboration.

REFERENCES

Adams, A.R., 2001. Strained quantum well lasers. IEEE J. Sel. Top. Quantum Electron. 17, 1364.

G. P. Agrawal and N. K. Dutta, 1993. Semiconductor Lasers. Springer. ISBN: 978-1-4612-7579-4.

Akahane, K., Yamamoto, N., Gozu, S.I., Ohtani, N., 2004. Heteroepitaxial growth of GaSb on Si (001) substrates. J. Cryst. Growth 264, 21.

Ayers, J.E., Showalter, L.J., Ghandhi, S.K., 1992. Post-growth thermal annealing of GaAs on Si(001) grown by organometallic vapor phase epitaxy. J. Cryst. Growth 125, 329–335.

Bahriz, M., Lollia, G., Baranov, A.N., Teissier, R., 2015. High temperature operation of far infrared (λ ≈ 20 μm) InAs/AlSb quantum cascade lasers with dielectric waveguide. Opt. Express 23, 1523.

Bai, Y., Slivken, S., Kuboya, S., Darvish, S.R., Razeghi, M., 2010. Quantum cascade lasers that emit more light than heat. Nat. Photonics 4, 99–102.

Baranov, A.N., Teissier, R., 2015. Quantum cascade lasers in the InAs/AlSb material system. IEEE J. Sel. Top. Quantum Electron. 21, 1200612.

Baranov, A.N., Bahriz, M., Teissier, R., 2016. Room temperature continuous wave operation of InAs-based quantum cascade lasers at 15 μm. Opt. Express 24, 18799.

Baski, A.A., Erwin, S.C., Whitman, L.J., 1997. The structure of silicon surfaces from (001) to (111). Surf. Sci. 392, 69.

Belenky, G., Shterengas, L., Kisin, M.V., Hosoda, T., 2013. Gallium antimonide (GaSb)-based type-I quantum well diode lasers: recent development and prospects. In: Baranov, A.N., Tournié, E. (Eds.), Semiconductor Lasers: Fundamental and Applications. Woodhead Publishing, pp. 441–486.

Bolkhovityanov, Y.B., Pchelyakov, O.P., 2008. GaAs epitaxy on Si substrates: modern status of research and engineering. Physics-Uspekhi 51, 437–456.

Borg, M., Schmid, H., Gooth, J., Rossell, M.D., Cutaia, D., Knoedler, M., Bologna, N., Wirths, S., Moselund, K.E., Riel, H., 2017. High-mobility GaSb nanostructures cointegrated with InAs on Si. ACS Nano 11, 2554–2560.

Cerba, T., Martin, M., Moeyaert, J., David, S., Rouviere, J.L., Cerutti, L., Alcotte, R., Rodriguez, J.B., Bawedin, M., Boutry, H., Bassani, F., Bogumilowicz, Y., Gergaud, P., Tournié, E., Baron, T., 2018. Anti-phase boundary free GaSb layer grown on 300mm (001)-Si substrate by metal organic chemical vapor deposition. Thin Solid Films 645, 5.

Cerutti, L., Rodriguez, J.B., Tournié, E., 2010. GaSb-based laser, monolithically grown on Si, emitting at 1.55 μm at room temperature. IEEE Photon. Technol. Lett. 22, 553–555.

Cerutti, L., Castellano, A., Rodriguez, J.-B., Bahri, M., Largeau, L., Balocchi, A., Madiomanana, K., Lelarge, F., Patriarche, G., Marie, X., Tournié, E., 2015. GaSb-based composite quantum wells for laser diodes operating in the telecom wavelength range near 1.55-μm. Appl. Phys. Lett. 106, 101102.

Choi, H.K., Hull, R., Ishiwara, H., Nemanich, R.J. (Eds.), 1988. Heteropitaxy on Si: Fundamentals, Structure, and Devices. In: vol. 116. Materials Research Society, Pittsburgh, PA.

Dier, O., Lin, C., Grau, M., Amann, M.-C., 2004. Selective and non-selective wet-chemical etchants for GaSb-based materials. Semicond. Sci. Technol. 19, 1250.

Foxon, C.T., Joyce, B.A., 1977. Interaction kinetics of As_2 and Ga on (100) GaAs surfaces. Surf. Sci. 64, 273.

Garbuzov, D.Z., Martinelli, R.U., Lee, H., York, P.K., Connolly, J.C., Narayan, S.Y., 1996. Ultralow-loss broadened-waveguide high-power 2 mm AlGaAsSb/InGaAsSb/GaSb separate-confinement quantum-well lasers. Appl. Phys. Lett. 69, 2006.

H. Habuka and T. Otsuka, Process for cleaning silicon semiconductor substrates, United States Patent 6,235,645 B1 (2001).

Huang, S.H., Balakrishnan, G., Khoshakhlagh, A., Jallipalli, A., Dawson, L.R., Huffaker, D.L., 2006. Strain relief by periodic misfit arrays for low defect density GaSb on GaAs. Appl. Phys. Lett. 88, 131911.
Huang, S.H., Balakrishnan, G., Khoshakhlagh, A., Dawson, L.R., Huffaker, D.L., 2008. Simultaneous interfacial misfit array formation and antiphase domain suppression on miscut silicon substrate. Appl. Phys. Lett. 93, 071102.
Huang, Y., Xiong, M., Wu, Q., Dong, X., Zhao, Y., Zhao, Y., Shi, W., Miao, X., Zhang, B., 2017. High-performance mid-wavelength InAs/GaSb superlattice infrared detectors grown by production-scale metalorganic chemical vapor deposition. IEEE J. Quantum Electron. 53 (5), 4000305.
Jallipalli, A., Kutty, M.N., Balakrishnan, G., Tatebayashi, J., Nuntawong, N., Huang, S.H., Dawson, L.R., Huffaker, D.L., Mi, Z., Bhattacharya, P., 2007. 1.54 μm GaSb/AlGaSb multi-quantum-well monolithic laser at 77 K grown on miscut Si substrate using interfacial misfit arrays. Electron. Lett. 43, 1198.
Jung, R., Herrick, J., Norman, K., Turnlund, C., Jan, K., Feng, A.C.G., Bowers, J.E., 2018. Impact of threading dislocation density on the lifetime of InAs quantum dot lasers on Si D. Appl. Phys. Lett. 112, 153507.
Kim, Y.H., Lee, J.Y., Noh, Y.G., Kim, M.D., Cho, S.M., Kwon, Y.J., Oh, J.E., 2006. Growth mode and structural characterization of GaSb on Si (001) substrate: a transmission electron microscopy study. Appl. Phys. Lett. 88, 241907.
Kroemer, H., 1987. Polar-on-non-polar epitaxy. J. Cryst. Growth 81, 193.
Lauer, C., Dier, O., Amann, M.-C., 2006. Low-resistive metal/n+-InAsSb/n-GaSb contacts. Semicond. Sci. Technol. 21, 1274.
Li, Q., Lau, K.M., 2017. For a recent review on these issues, epitaxial growth of highly mismatched III–V materials on (001) silicon for electronics and optoelectronics. Prog. Cryst. Growth Charact. Mater. 63, 105.
Liu, P.Q., Hoffman, A.J., Escarra, A.D., Franz, K.J., Khurgin, J.B., Dikmelik, Y., Wang, X., Fan, J.-Y., Gmachl, C.F., 2010. Highly power-efficient quantum cascade lasers. Nat. Photonics 4, 95–98.
Madiomanana, K., Bahri, M., Rodriguez, J.B., Largeau, L., Cerutti, L., Mauguin, O., Castellano, A., Patriarche, G., Tournié, E., 2015. Most relevant references can be found in, e.g., silicon surface preparation for III–V molecular beam epitaxy. J. Cryst. Growth 413, 17.
Majtykaa, A., Trębalaa, M., Tukiainen, A., Chrobak, D., Borgieła, W., Räisänen, J., Nowak, R., 2016. Elastic–plastic transition in MBE-grown gasb semiconducting crystal examined by nanoindentation. Acta Phys. Polon. A 130, 1131.
Martin, M., Caliste, D., Cipro, R., Alcotte, R., Moeyaert, J., David, S., Bassani, F., Cerba, T., Bogumilowicz, Y., Sanchez, E., Ye, Z., Bao, X.Y., Pin, J.B., Baron, T., Pochet, P., 2016. Toward the III–V/Si co-integration by controlling the biatomic steps on hydrogenated Si(001). Appl. Phys. Lett. 109, 253103.
Nguyen-Van, H., Baranov, A.N., Loghmari, Z., Cerutti, L., Rodriguez, J.-B., Tournet, J., Narcy, G., Boissier, G., Patriarche, G., Bahriz, M., Tournié, E., Teissier, R., 2018. Quantum cascade lasers grown on silicon. Sci. Rep. 8, 7206.
Onabe, K., 1982. Unstable regions in III–V quaternary solid solutions composition plane calculated with strictly regular solution approximation. Jpn. J. Appl. Phys. 21, L323.
Razeghi, M., et al., 2015. Quantum cascade lasers: from tool to product. Opt. Express 23, 8462–8475.
Reboul, J.R., Cerutti, L., Rodriguez, J.B., Grech, P., Tournié, E., 2011. Continuous-wave operation above room temperature of GaSb-based laser diodes grown on Si. Appl. Phys. Lett. 99, 121113.
Richardson, C.J.K., He, L., Kanakaraju, S., 2011. Metamorphic growth of III–V semiconductor bicrystals. J. Vac. Sci. Technol. B 29, 03C126.

Rocher, A., Snoeck, E., 1999. Misfit dislocations in (001) semiconductor heterostructures grown by epitaxy. Mater. Sci. Eng. B 67, 62.

Rodriguez, J.B., Cerutti, L., Grech, P., Tournié, E., 2009. Room-temperature operation of a 2.25 μm electrically pumped laser fabricated on a silicon substrate. Appl. Phys. Lett. 94, 061124.

Rodriguez, J.-B., Madiomanana, K., Cerutti, L., Castellano, A., Tournié, E., 2016. X-ray diffraction study of GaSb grown by molecular beam epitaxy on silicon substrates. J. Cryst. Growth 439, 33–39.

Rogalski, A., Martyniuk, P., Kopytko, M., 2017. InAs/GaSb type-II superlattice infrared detectors: future prospect. Appl. Phys. Rev. 4, 031304.

Rösch, M., Scalari, G., Beck, M., Faist, J., 2015. Octave-spanning semiconductor laser. Nat. Photonics 9, 42–47.

Rothman, L.S., Gordon, I.E., Babikov, Y., Barbe, A., Chris Benner, D., Bernath, P.F., et al., 2013. The HITRAN2012 molecular spectroscopic database. J. Quant. Spectrosc. Radiat. Transf. 130, 4–50.

Rouillard, Y., Lambert, B., Toudic, Y., Baudet, M., Gauneau, M., 1995. On the use of dimeric antimony in molecular beam epitaxy. J. Cryst. Growth 156, 30.

Salhi, A., Rouillard, Y., Angellier, J., Garcia, M., 2004. Very-low-threshold 2.4-μm GaInAsSb–AlGaAsSb laser diodes operating at room temperature in the continuous-wave regime. IEEE Photon. Technol. Lett. 16, 2424.

Shin, K.W., Kim, H.-W., Kim, J., Yang, C., Lee, S., Yoon, E., 2010. The effect of low temperature buffer layer on the growth of pure Ge on Si(001). Thin Solid Films 518, 6496–6499.

Sirtori, C., Teissier, R., 2010. In: Paiella, R. (Ed.), Intersubband Transitions in Quantum Structures. McGraw-Hill Professional Publishing.

Steenbergen, E.H., Ariyawansa, G., Reyner, C.J., Jenkins, G.D., Morath, C.P., Duran, J.M., Scheihing, J.E., Cowan, V.M., 2017. A recent review of mid-wavelength infrared type-II superlattices: carrier localization, device performance, and radiation tolerance. Proc. SPIE 10111, 1011104.

Stringfellow, G.B., 1982. Miscibility gaps in quaternary III–V alloys. J. Cryst. Growth 58, 194.

Tournié, E., Baranov, A.N., 2012. Mid-infrared lasers: a review. In: Coleman, J.J., Brice, A.C., Jagadish, C. (Eds.), Advances in Semiconductor Lasers. In: Semiconductors and Semimetals, vol. 86. Academic Press, pp. 183–226.

Trampert, A., Tournié, E., Ploog, K.H., 1995. Novel plastic strain-relaxation mode in highly mismatched III–V layers induced by two dimensional epitaxial growth. Appl. Phys. Lett. 66, 2265.

van der Ziel, J.P., Malik, R.J., Walker, J.F., Mikulyak, R.M., 1986. Optically pumped laser oscillation in the 1.6-1 8 μm region from strained layer $Al_{0.4}Ga_{0.6}Sb/GaSb/Al_{0.4}Ga_{0.6}Sb$ double heterostructures grown by molecular beam hetero-epitaxy on Si substrate. IEEE J. Quantum Electron. 22, 1587.

Volz, K., Beyer, A., Witte, W., Ohlmann, J., Németh, I., Kunert, B., Stolz, W., 2011. GaP-nucleation on exact Si (0 0 1) substrates for III/V device integration. J. Cryst. Growth 315, 37.

Vurgaftman, I., Meyer, J.R., Ram-Mohan, L.R., 2001. Band parameters for III–V compound semiconductors and their alloys. J. Appl. Phys. 89, 5815.

Vurgaftman, I., Weih, R., Kamp, M., Meyer, J.R., Canedy, C.L., Kim, C.S., Kim, M., Bewley, W.W., Merritt, C.D., Abell, J., Höfling, S., 2015. Interband cascade lasers. J. Phys. D 48, 123001.

Wang, C.A., 2004. Progress and continuing challenges in GaSb-based III–V alloys and heterostructures grown by organometallic vapor-phase epitaxy. J. Cryst. Growth 272, 664.

Wu, H., Zhang, Y.Y., Razeghi, M., 2018. Room temperature operation of InxGa12xSb/InAs type-II quantum well infrared photodetectors grown by MOCVDD. Appl. Phys. Lett. 112, 111103.
Yamaguchi, M., Tachikawa, M., Itoh, Y., Sugo, M., Kondo, S., 1990. Thermal annealing effects of defect reduction in GaAs on Si substrates. J. Appl. Phys. 68, 4518.
Yao, Y., Hofman, A.J., Gmachl, C.F., 2012. Mid-infrared quantum cascade lasers. Nat. Photonics 6, 432–439.

CHAPTER TWO

III–V on Silicon Nanocomposites

Johann Peter Reithmaier[1], Mohamed Benyoucef
Technische Physik, Institute of Nanostructure Technologies and Analytics (INA), Center of Interdisciplinary Nanostructure Science and Technology (CINSaT), University of Kassel, Kassel, Germany
[1]Corresponding author: e-mail address: jpreith@ina.uni-kassel.de

Contents

1. INTRODUCTION

Silicon is the dominating semiconductor in the electronics industry and may become also the most important material platform for photonic integration in future. There is a long-lasting dream to combine passive photonics with active optoelectronics on a common silicon-based platform. However, due to the indirect bandgap of this material no well-performing active device could be realized. In contrast to that, the highest performing optoelectronic devices are realized on III–V compound semiconductor heterostructures. The integration of III–V materials on silicon chips is sought as one of the major solutions to combine active light sources with silicon photonics and electronics (Chen et al., 2011; Egorov et al., 1999; Heitz et al., 1999; Kim et al., 2010; Ko et al., 2010; Monat et al., 2002; Palit et al., 2009; Seassal et al., 2001; Tanabe et al., 2010; Yoon et al., 2010). Also, since many

Semiconductors and Semimetals, Volume 99
ISSN 0080-8784
https://doi.org/10.1016/bs.semsem.2018.08.004

years, III–V integration is seen as a promising approach for low-cost high-speed communication and computing technologies (Goodman et al., 1984; Miller, 2000).

First approaches to realize optical transitions utilize SiGe/Si short-period superlattices (Abstreiter, 1993) or defect atoms (e.g., rare earth atoms) (Castagna et al., 2003).

Other interesting epitaxial approaches are based on Ga(N,As,P) compounds, which can be grown lattice matched to Si (Liebich et al., 2011), or the growth of new Si-based compounds, such as SiGeSn, which allow in certain conditions the realization of a direct bandgap. In spite of severe material quality problems, such as strong segregation effects and very low growth temperatures, first low-temperature electroluminescence and lasing could be obtained (Schwartz et al., 2015; Wirths et al., 2015).

More applicable for heterointegration are chip or wafer bonding techniques (Duan et al., 2014, 2016; Heck et al., 2011; Lamponi et al., 2012). However, wafer scale bonding is not very economic because the majority of III–V material has to be removed and is wasted. Monolithic integration by direct epitaxial growth on Si avoids the substrate removal but has to accommodate the large lattice mismatch and the different expansion coefficients. For this approach, different techniques were used to relax the strain to the appropriate III–V lattice constants, such as Si/SiGe graded substrates (Tanoto et al., 2009), Ge-on-insulator on Si (Bordel et al., 2010), or thick III–V relaxation layers (Gérard et al., 1996; Linder et al., 1999; Luxmoore et al., 2013; Mi et al., 2009; Wang et al., 2011). Although thick buffer layers increase the process complexity, several successful realizations of 1.3 μm quantum dot (QD) lasers were reported (Liu et al., 2011, 2014, 2015).

A possibility to reduce significantly the density of remaining threading dislocations during direct III–V growth on Si is selective area growth and the geometric restriction in nanostructures (Lourdudoss, 2012; Wang et al., 2015). High material quality can be obtained but only on limited areas. All above-mentioned techniques have a fundamental drawback, which is the need of separate fabrication equipments for III–V materials and silicon due to cross-contamination. In case of wafer bonding, the waste of the expensive III–V material has to be added. Therefore, none of the above-discussed approaches are ideal and fulfill the demand on a new optoelectronic material compatible with silicon processing.

An ideal material should be processed like Si or Si-compatible materials without cross-contamination and still have optoelectronic properties similar to III–V direct bandgap material. In the last years, a new approach based on a III–V–Si nanocomposite was investigated, which is from the basic point of view envisioning such an ideal composite material. This chapter reviews some of the major initial results gained within the last few years toward a full optoelectronic silicon-based integration platform. This approach is based on the growth of III–V QDs directly on either planar silicon (Al-Zoubi et al., 2011; Benyoucef et al., 2013) or patterned surfaces (Benyoucef et al., 2012; Benyoucef and Reithmaier, 2013; Usman et al., 2015). High optical quality from single QDs directly grown on Si surfaces was obtained using GaAs/InAs core–shell structures (Benyoucef et al., 2013). A defect-free single crystal planar silicon layer with embedded InAs nanoclusters was realized (Benyoucef et al., 2014), which was confirmed by detailed transmission electron microscopy studies showing fully relaxed nanoclusters. The relaxation takes place by closed loop interface dislocations without initiating threading dislocations in the silicon matrix (Wu et al., 2015).

2. VISION OF A NEW SI-BASED NANOCOMPOSITE MATERIAL

The nanocomposite material consisting of III–V QDs fully embedded in Si and can be processed like Si. In this case, the optoelectronic material is minimized to the lowest volume necessary to form a direct band structure and to confine electrons locally in a III–V heterostructure. In combination with either selective area growth or planar growth and subsequent etching, the active material can be localized. If the optical gain is large enough, i.e., in the order of at least 10–20 cm^{-1}, one can realize multifunctional optoelectronic devices. In Fig. 1, a sketch of a possible Si-photonic chip with integrated nanocomposite material is illustrated consisting of passive and active components including light emitter (laser), modulator (EAM: electroabsorption modulator), amplifier (SOA: semiconductor optical amplifier), and detector.

Such a new material would have several advantages:

- III–V material is minimized and is only localized at active devices.
- No III–V material is accessible from the surface, i.e., standard silicon processing is possible without cross-contamination.
- No extra III–V processing is needed except the epitaxy step for the nanocomposite material.

However, the realization might be still very challenging and the following issues have to be addressed:

- Light emission has to be below the Si bandgap in the wavelength range of 1.2–1.7 μm.
- Silicon overgrowth on III–V material has to be established, which allow single crystal quality and a planar surface.
- The large differences in growth temperatures between InAs (typ. 450°C) and Si (typ. 800°C) have to be accommodated.
- The high lattice mismatch between III–V material and Si has to be adjusted.
- The defect generation has to be minimized to allow efficient radiative recombination.

A possible way to accommodate most of these requests is illustrated in Fig. 2. Here, nanoheteroclusters based on an InAs/GaAs core–shell geometry are

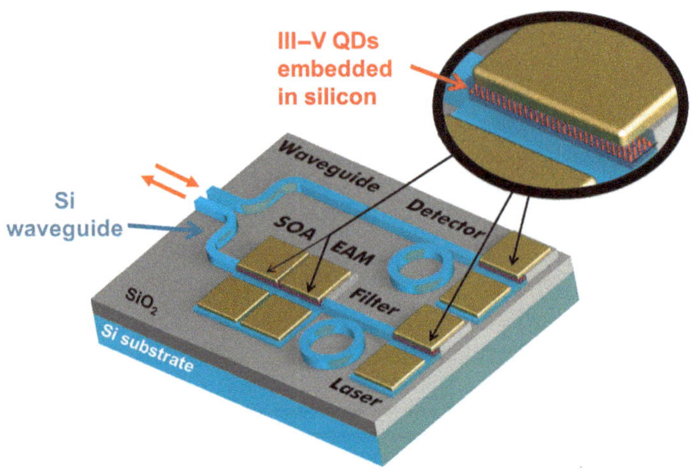

Fig. 1 Sketch of monolithic integration of active and passive photonic devices on a newly proposed III–V/nSi platform with embedded III–V QDs.

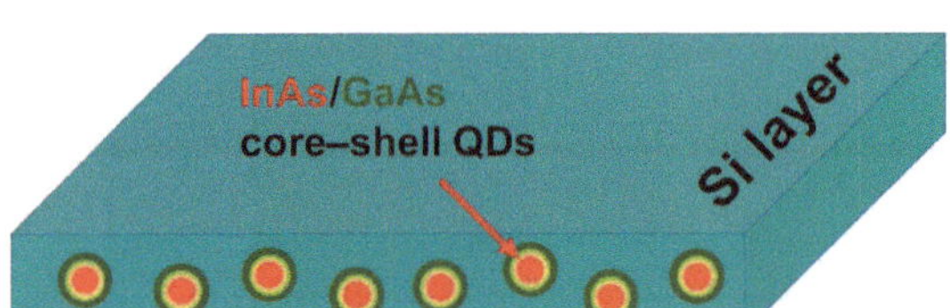

Fig. 2 Illustration of active nanocomposite material with embedded core–shell QDs in planar crystalline silicon.

used to confine charge carriers in the III–V material. This avoids a significant overlap with defects at the interface between III–V and Si. The use of nanoclusters allows also a direct access to the single crystal silicon surface between the InAs islands during the Si overgrowth. This enables a coherent overgrowth to form monolithic Si.

In the following, different initial results will be presented and discussed toward this vision of a new nanocomposite optically active material.

3. DIRECT EPITAXY OF III–V QDs ON Si

3.1 InAs QD Growth on Silicon

All samples discussed in this section were grown on exactly oriented n-type (100) silicon substrates. The growth was performed by a GEN II Varian molecular beam epitaxy (MBE) system. This III–V MBE system was modified by installing an additional silicon e-beam evaporator and a high-temperature silicon substrate heater to allow silicon homoepitaxy within a III–V environment. With this modified equipment, substrate temperatures up to 1200°C and growth rates in the range of 200–300 nm/h are possible. The substrates were ex situ cleaned using buffered HF (NH_4/HF: H_2O) (1:1) for 4 min as a preremoval step of the surface native oxide followed by an in situ thermal atomic hydrogen-assisted surface cleaning process at 500°C (45 min with $P_H = 3.7 \times 10^{-7}$ Torr). The oxide was removed by a thermal process performed in the growth chamber using a temperature of 910°C. The surface quality and the dots formation process are monitored in situ by reflection high-energy diffraction. Additional improvement for the cleaning and growth was achieved by exposing the silicon surface with a low Ga flux (0.1 ML/s). The influences of different growth parameters, such as growth temperature, V/III ratio, and In-growth rate on the structural properties of InAs QDs, are investigated.

In Fig. 3, the influence of the III/V ratio (A–C) and the substrate temperature (D–F) are shown. With increasing III–V ratio, the migration length of In atoms is strongly suppressed and reaches an optimum at about 1:25 (=beam equivalent pressure ratio) as seen in Fig. 3B. Further increase results in the formation of clusters of QDs. With increasing temperature, the In desorption play a more important role. This leads to a reduction in dot density (see Fig. 3E) and finally avoids any dot formation at 500°C (see Fig. 3F). An optimum condition was concluded to be at 400°C growth temperature with a V–III ratio of 25. This results in a high dot density of about 10^{11} cm^{-2} and a dot size of about 35 nm in width and 6.5 nm in height.

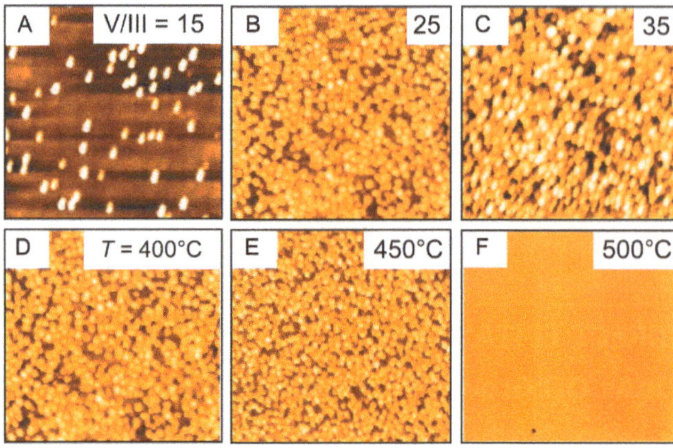

Fig. 3 (A–C) InAs QDs formation at different V/III ratios. Growth temperature was at 400°C. (D–F) InAs QDs formation at different growth temperatures (measured by pyrometer). The V/III ratio is 25 (Benyoucef et al., 2012).

3.2 Position-Controlled QDs Formation

For the control of the position of the QD formation, prepatterned Si surfaces were prepared by electron beam lithography and dry chemical etching. Details about the structuring process of nanoholes are in Benyoucef et al. (2012) and Usman et al. (2015).

After the resist removal using a sequence of cleaning steps involving acetone and isopropyl alcohol, the substrate was then further cleaned in oxygen plasma asher for 30 min to remove any residual resist particles and carbon contamination left after the lithographic processing. Finally the substrate was cleaned with HF:H_2O (1:2) solution for 2 min to remove the native oxide. The substrates were then loaded into MBE system within 10 min of ex situ chemical cleaning to perform in situ surface preparation and subsequent growth process.

Following this approach, the prepatterned sample was exposed to AsH_3 flux of 3×10^{-7} Torr at 500°C for 45 min then the sample was annealed at 750°C for 30 min to desorb the native oxide from the surface and the holes. Schematic sequence of the growth on patterned sample is shown in Fig. 4 (top left). The grown sample consists of 2 nm nominal thickness of a GaAs buffer at 600°C, 2 MLs nominal thickness of an $In_{0.15}Ga_{0.85}As$ QDs layer at 500°C, and 2 nm nominal thickness of a GaAs capping layer at 500°C. The growth of GaAs capping layer at temperature of 500°C same as that for QDs growth was intended to avoid high indium desorption.

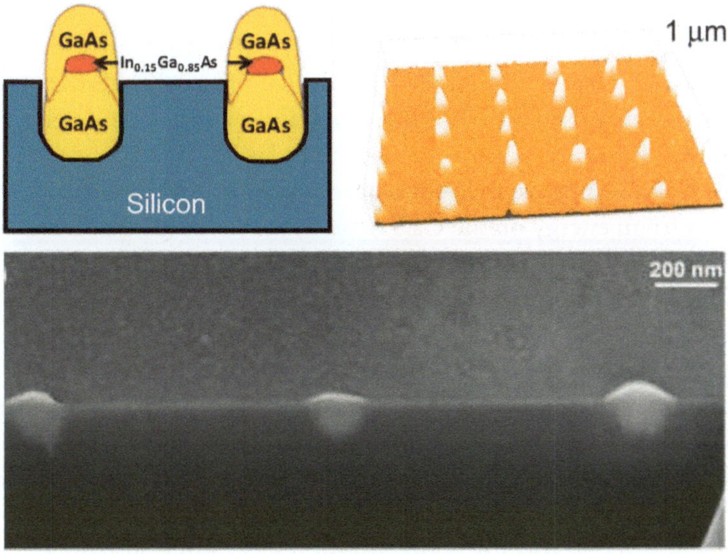

Fig. 4 (*Top, left*) Schematic of selectively grown core–shell InGaAs/GaAs QDs in prepatterned holes. (*Top, right*) AFM image ($5 \times 5\ \mu m^2$) of optimized GaAs grown on prepatterned silicon substrate with a hole array of 1 μm in period. (*Bottom*) A crosscut made by a FIB through the etched holes filled with III–V QDs (Benyoucef et al., 2012).

A 3D AFM image of ordered arrays of GaAs nanoislands is shown in Fig. 4 (top right). Note that the nucleation of the GaAs takes place in all the nanoholes. This is a strong indication that the oxide is removed from all the nanoholes. The growth at 600°C also eliminates unnecessary nucleation of islands between the holes due to longer Ga adatom migration length and enhanced desorption rate. An SEM cross-sectional image (tilted by 54 degree from normal) of the GaAs islands, obtained after cutting through the islands with the help of a focused ion beam (FIB), is provided in Fig. 4 (bottom). The cross-sectional image confirms highly selective formation of GaAs islands inside the patterned holes as well as a good interface with the bottom and sidewalls of the holes.

3.3 InAs/GaAs Core–Shell QDs Directly Grown on Silicon

Directly grown InAs QDs are suffering by nonsaturated dangling bonds and a high density of defects at the Si–InAs interface. Owing to the large lattice mismatch of 11%, about every 10th atom will be a defected atom to accommodate the strain. Therefore, excited electron–hole pairs will be quickly captured by these defects and will recombine nonradiatively.

To avoid this direct access to recombination centers, core–shell QDs are formed by a three-step growth process. First, GaAs islands are formed in the Stranski–Krastanov growth mode on 5 degree off-cut Si substrate. An off-cut substrate was used to avoid the formation of antiphase defects, which cannot be suppressed without a thick buffer layer.

The GaAs islands are then subsequently overgrown by InAs. Owing to the lower strain energy on the GaAs islands, the InAs QD formation is preferentially taking place on top of the GaAs islands. In a final last step, these GaAs/InAs islands were overgrown by GaAs again. Details about the growth process are reported in Benyoucef et al. (2013). In Fig. 5 (top left), an SEM image of an example for a GaAs/InAs core–shell QD is shown. Clear faceting can be seen. The QD size is relatively large with lateral dimension of more than 100 nm and a height in the order of 20–30 nm. However, due to significant intermixing between InAs and GaAs, the emission wavelength is in the order of 940–950 nm at 5 K. In Fig. 5 (right), low-temperature (5 K) single-QD narrow linewidth emission spectrum is shown with a clear signature of an exciton (x)–biexciton (xx) behavior (see inset of Fig. 5).

This example shows that QDs with high optical quality can be directly grown on Si surfaces using a core–shell geometry, which confines the carriers within an inner defect-free III–V heterostructure and avoids successfully nonradiative recombination by the dense defect network at the III–V to Si interface.

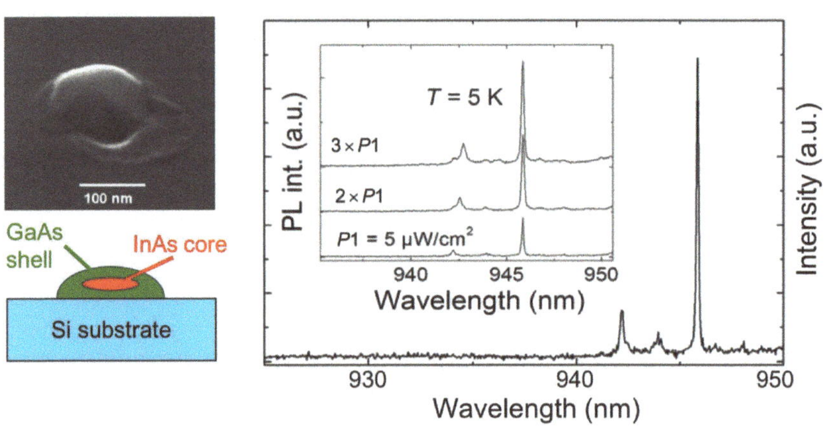

Fig. 5 (*Top, left*) SEM image of a large core–shell InGaAs/GaAs QD grown on Si (5 degree off-cut). (*Bottom left*) Illustration of a core–shell QD on Si substrate. (*Right*) Low-temperature PL spectra. The *inset* shows the spectra at different excitation power densities clearly indicating X and XX exciton lines (Benyoucef et al., 2013).

4. MONOLITHIC INTEGRATION OF III–V NANOCLUSTERS INTO A SI MATRIX

The vision is to realize a fully embedded III–V material in silicon and creating a hybrid material that can be structured only by silicon processing. To realize such a material, a multistep process was developed, which allows to combine a single crystal high-temperature growth process for silicon with a low-temperature deposition process of III–V material, such as InAs. Details are reported in Benyoucef et al. (2014).

4.1 Multistep Growth Process

Before the growth starts, the same ex situ and in situ cleaning process is performed as mentioned earlier including a thermal oxide removal in the growth chamber. The first epitaxial step is the establishment of a well-defined silicon surface by the homoepitaxy growth of a silicon layer, which took place at 700°C and a growth rate of about 200 nm/h. In comparison to previous works where mainly Si doping cells were used, this system configuration allows several orders of magnitude lower background contamination due to the much higher growth rates. To get a high-quality Si crystal, an additional postgrowth annealing (PGA) step was applied for 10 min.

After the growth of 50 nm silicon buffer layer at 700°C followed by a PGA step at 800°C for 10 min, then the substrate temperature was reduced to 400°C for the deposition of typically two monolayers (MLs) of InAs. This process step, as illustrated in Fig. 6A, results in a high density of InAs islands without closed wetting layer. Therefore, the growth mode can be more classified as Volmer–Weber than Stranski–Krastanov. This is also of central importance for the subsequent Si overgrowth, because it allows a direct access to the silicon surface and preserves the coherence of the crystal. Owing to the high strain, a dense defect network can be seen in the TEM picture.

As illustrated in Fig. 6B, after overgrowth with a 10 nm thick silicon layer at the same growth temperature of InAs (of 400°C), the surface has many nanoholes with diameters on the scale of typical QD diameters. Owing to the low mobility of Si atoms at this temperature and the direct access to the Si crystal, growth took place only in the surrounding of the InAs islands. As a consequence, nanoholes were formed with an InAs QD in the hole as can be seen in the corresponding TEM picture.

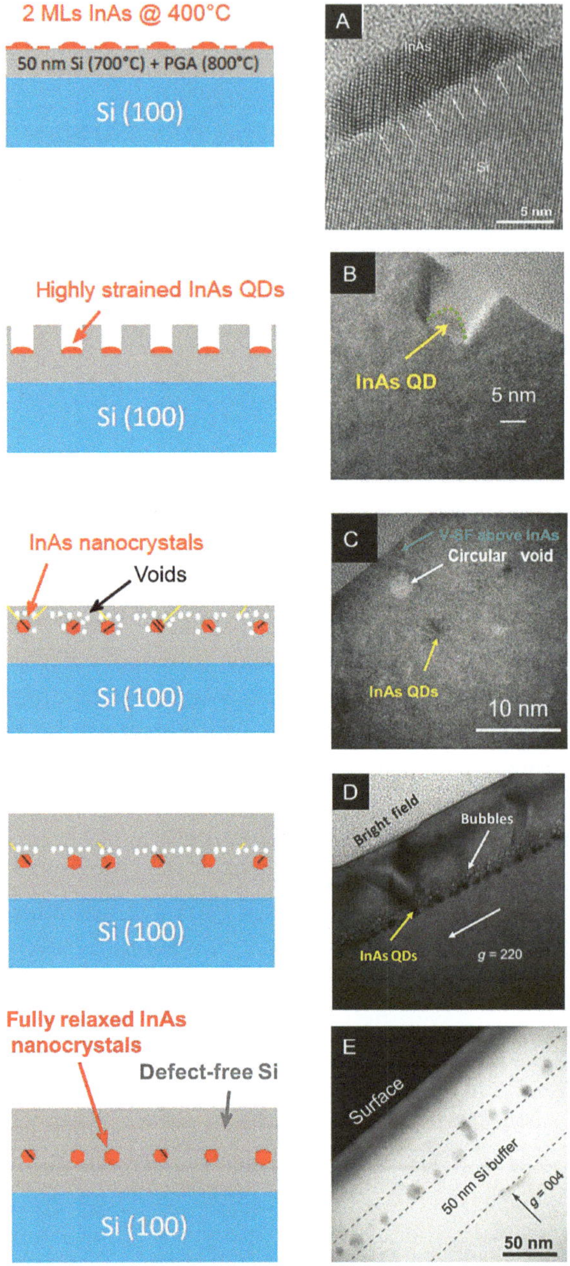

Fig. 6 Process sequence used to embed InAs clusters into a silicon matrix with schematic illustration in left column and the corresponding TEM image in right column. (A) Initial InAs QD growth (400°C) on a Si buffer layer, which was grown by MBE at 700°C and postgrowth annealing (PGA) at 800°C. The TEM image shows a single lens shape InAs QD on Si surface with a dense dislocation network. (B) Si overgrowth at 400°C (10 nm). InAs QDs acts as a selective epitaxial mask for Si and forms nanoholes. (C) Annealing step at 700°C. Either certain closed voids are formed or InAs nanocrystals are fully embedded in Si. Aligned to the original holes and underlying InAs QDs, vacancy-stacking faults (V-SF) are observed. (D) Si overgrowth at 700°C (40 nm). Voids are reduced. (E) Final additional annealing step at 800°C. Clusters of different sizes are formed without voids and the defects are strongly decreased (Benyoucef et al., 2014).

To close the holes, another PGA step at 700°C was performed. As can be seen in Fig. 6C, this step was only partially successful. Although the holes were closed, there are still voids remaining, which are indicated in TEM picture. Here, a rearrangement of the InAs takes place and from a lens-shaped QD on the surface and fully embedded spherical InAs clusters are formed. The Moiré pattern indicates the large lattice mismatch between Si and InAs and will be addressed later.

As a next step, 40 nm Si was overgrown at 700°C. This step further reduces the void density, however, as illustrated in Fig. 6D, cannot completely remove them. Therefore, a third PGA step at 800°C is needed to remove all voids and to recrystallize the InAs clusters to nearly dislocation-free QDs. This final step is illustrated in Fig. 6E. The corresponding TEM picture shows that InAs QDs are located at the first 10 nm of Si overgrow layer (indicated by dashed lines). The top 40 nm Si overgrow layer does not show any misfit dislocations anymore.

4.2 Detailed TEM Investigations

The crystal quality and other crystal properties, such as strain distribution and type of defects, are investigated by high-resolution TEM (HRTEM). Details are reported in Benyoucef et al. (2014) and Wu et al. (2015). In Fig. 7A, a cross-sectional TEM image is shown after the full multistep process as described earlier. For AFM investigations, additional InAs islands are deposited on the surface, which can be seen as lens-shaped structures in Fig. 7A. Owing to the 11% lattice mismatch between Si with high monocrystalline quality and the fully relaxed InAs clusters, Moiré patterns can be

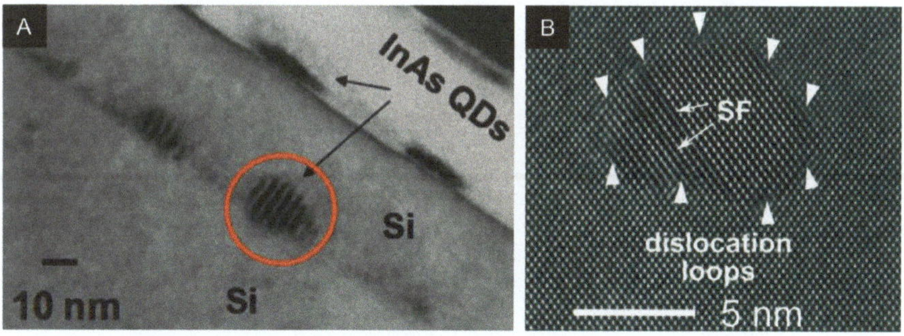

Fig. 7 (A) TEM image of InAs islands on the Si surface and fully embedded InAs clusters. The Moiré patterns are caused by the lattice difference between III–V material and Si. (B) An atomic resolution image of an InAs cluster exhibiting stacking faults (SF) and a surrounding dislocation network at the III–V to Si interface. The embedding Si crystal is defect free (Benyoucef et al., 2014).

seen as electron interference patterns. The marked InAs cluster in Fig. 7A, show a parallel interference pattern, which clearly indicates a defect-free III–V material. In Fig. 7B, a HRTEM image of an InAs nanocluster is shown with two stacking faults, as indicated in the picture. However, the relaxation has taken place by dislocations formed at the III–V to Si interface. Here, dislocation loops are formed (indicated by triangles), which do not penetrate into the embedding Si matrix. Therefore, the overgrown and surrounding Si is completely defect free and single crystal.

The full relaxation of the InAs clusters was approved by strain analysis. In Fig. 8, two fully embedded InAs clusters are shown. In (A), a cluster is selected, which shows in the bulk-like area of the InAs cluster no stacking fault or other defects. The defects are pinned at the interface as dislocation loops. The corresponding strain field component ε_{xx} is plotted on the right side (Fig. 8B) and shows a very homogeneously relaxed situation with an average strain value well below 1% in comparison to 11.6% in the unrelaxed case. In Fig. 8C and D, another example is shown, which exhibits a single stacking fault. Also here, the relaxation takes place at the interface and no dislocations are seen in the embedding Si.

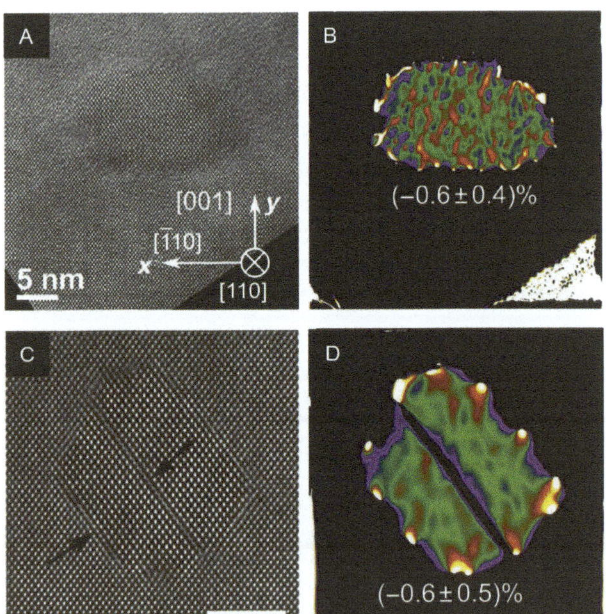

Fig. 8 Strain analysis of two examples of InAs clusters using HRTEM images (A) and (C) and extracted material strain component ε_{xx}. (B) and (D). In (A) and (B), an InAs cluster is shown without any defects in the bulk region of the cluster. In (C) and (D), the InAs cluster shows one clear stacking fault (Wu et al., 2015).

5. CONCLUSIONS

A brief overview is given about different approaches for the heterogeneous as well as monolithic integration of III–V materials on silicon for integrated photonics. While bonding techniques based on a flip-chip approach or molecular bonding on a die level are already mature and used in production, full wafer bonding, and monolithic integration by direct growth on silicon needs to be qualified and reliability issues have to be solved. New approaches are coming up based on selective area growth or lattice matched III–V compounds as well as band structure engineered new materials, such as SiGeSn, to allow full wafer processing. However, all of those approaches still need separate Si and III–V processing equipment and are only partially compatible with CMOS processing.

Finally, a fundamentally different approach is presented here based on a new silicon-like bulk material but with fully embedded III–V nanoclusters, which may allow in future combining Si-like processing behavior with optical direct bandgap features. Major steps toward this goal were shown, such as homogenously relaxed InAs nanoclusters fully embedded in defect free, planar Si matrix, and the demonstration of GaAs/InAs core–shell QDs directly grown on Si with bright and narrow line single dot optical emission.

Although the results show very promising steps toward an ultimate new optoelectronic Si-based material, there are still several challenges to be solved, such as the embedding of optically active core–shell QDs into a Si matrix, to realize high enough optical gain ($>10\ \mathrm{cm}^{-1}$) for integrated devices, to understand the band alignment at the III–V to Si interface, to demonstrate electrically driven optical gain, and to be aware of possible III–V diffusion properties at high temperatures. However, we believe that those basic issues can be solved in the near future, and hence, the concept has the potential to substitute finally all other approaches due to the highest CMOS process compatibility, minimal utilization of the III–V material, full wafer processing, and flexibility in device processing.

ACKNOWLEDGMENTS

We acknowledge M. Usman for the growth of the samples and the processing. T. Al-Zoubi for the growth. M. Wu, A. Trampert from Paul Drude Institute in Berlin for TEM measurements. We would like to thank the German Federal Ministry of Education and Research for the financial support by the project "Monalisa."

REFERENCES

Abstreiter, G., 1993. Physics and perspectives of Si/Ge heterostructures and superlattices. Phys. Scr. T49, 42–45.

Al-Zoubi, T., Usman, M., Benyoucef, M., Reithmaier, J.P., 2011. Growth of InAs quantum dots and dashes on silicon substrates: formation and characterization. J. Cryst. Growth 323, 422–425.

Benyoucef, M., Reithmaier, J.P., 2013. Direct growth of III–V quantum dots on silicon substrates: structural and optical properties. Semicond. Sci. Technol. 28, 094004.

Benyoucef, M., Usman, M., Al-Zoubi, T., Reithmaier, J.P., 2012. Pre-patterned silicon substrates for the growth of III–V nanostructures. Phys. Status Solidi A 209, 2402–2410.

Benyoucef, M., Usman, M., Reithmaier, J.P., 2013. Bright light emissions with narrow spectral linewidths from single InAs/GaAs quantum dots directly grown on silicon substrates. Appl. Phys. Lett. 102, 132101.

Benyoucef, M., Al-Zoubi, T., Reithmaier, J.P., Wu, M., Trampert, A., 2014. Nanostructured hybrid material based on highly mismatched III–V nanocrystals fully embedded in silicon. Phys. Status Solidi A 211, 817–822.

Bordel, D., Guimard, D., Rajesh, M., Nishioka, M., Augendre, E., Clavelier, L., Arakawa, Y., 2010. Growth of InAs/GaAs quantum dots on germanium-on-insulator-on-silicon (GeOI) substrate with high optical quality at room temperature in the 1.3 μm band. Appl. Phys. Lett. 96, 043101.

Castagna, M., Coffa, S., Monaco, M., Caristia, L., Messina, A., Mangano, R., Bongiorno, C., 2003. Si-based materials and devices for light emission in silicon. Phys. E 16, 547–553.

Chen, R., Tran, T.T.D., Ng, K.W., Ko, W.S., Chuang, L.C., Sedgwick, F.G., Chang-Hasnain, C., 2011. Nanolasers grown on silicon. Nat. Photonics 5, 170–175.

Duan, G.-H., Jany, C., Le Liepvre, A., Accard, A., Lamponi, M., Make, D., Kaspar, P., 2014. Hybrid III–V on silicon lasers for photonic integrated circuits on silicon. Proc. SPIE 9002, 90020X. https://doi.org/10.1117/12.2044258.

Duan, G.-H., Olivier, S., Jany, C., Malhouitre, S., Le Liepvre, A., Shen, A., Pommarede, X., Levaufre, G., Girard, N., Make, D., Glastre, G., Decobert, J., Lelarge, F., Brenot, R., Charbonnier, B., 2016. Hybrid III–V silicon photonic integrated circuits for optical communication applications. IEEE J. Sel. Top. Quantum Electron. 22, 6100711.

Egorov, A.Y., Kovsh, A.R., Ustinov, V.M., Zhukov, A.E., Maksimov, M.V., Cirlin, G.E., Ledentsov, N.N., Bimberg, D., Werner, P., Alferov, Z.I., 1999. Self-organized InAs quantum dots in a silicon matrix. J. Cryst. Growth 201/202, 1202–1204.

Gérard, J.M., Cabrol, O., Sermage, B., 1996. InAs quantum boxes: highly efficient radiative traps for light emitting devices on Si. Appl. Phys. Lett. 68, 3123.

Goodman, J.W., Leonberger, F.J., Kung, S.Y., Athale, R.A., 1984. Optical interconnections for VLSI systems. Proc. IEEE 72, 850–866.

Heck, M., Chen, H., Fang, A., Koch, B., Liang, D., Park, H., Sysak, M., Bowers, J., 2011. Hybrid silicon photonics for optical interconnects. IEEE J. Sel. Top. Quantum Electron. 17, 333–346.

Heitz, R., et al., 1999. Optical properties of InAs quantum dots in a Si matrix. Appl. Phys. Lett. 74, 1701–1703.

Kim, R.-H., et al., 2010. Waterproof AlInGaP optoelectronics on stretchable substrates with applications in biomedicine and robotics. Nat. Mater. 9, 929–937.

Ko, H., et al., 2010. Ultrathin compound semiconductor on insulator layers for high-performance nanoscale transistors. Nature 468, 286–289.

Lamponi, M., Keyvaninia, S., Jany, C., Poingt, F., Lelarge, F., de Valicourt, G., Roelkens, G., 2012. Low-threshold heterogeneously integrated InP/SOI laser with a double adiabatic taper coupler. IEEE Photon. Technol. Lett. 24, 76–78.

Liebich, S., Zimprich, M., Beyer, A., Lange, C., Franzbach, D.J., Chatterjee, S., Hossain, N., Sweeney, S.J., Volz, K., Kunert, B., Stolz, W., 2011. Laser operation of Ga(NAsP) lattice-matched to (001) silicon substrate. Appl. Phys. Lett. 99, 071109.

Linder, K.K., Phillips, J., Qasaimeh, O., Liu, X.F., Krishna, S., Bhattacharya, P., Jiang, J.C., 1999. Self-organized In0.4Ga0.6As quantum-dot lasers grown on Si substrates. Appl. Phys. Lett. 74, 1355.

Liu, H., Wang, T., Jiang, Q., Hogg, R., Tutu, F., Pozzi, F., Seeds, A., 2011. Long-wavelength InAs/GaAs quantum-dot laser diode monolithically grown on Ge substrate. Nat. Photonics 5, 416.

Liu, A.Y., Zhang, C., Norman, J., Snyder, A., Lubyshev, D., Fastenau, J.M., Liu, A.W.K., Gossard, A.C., Bowers, J.E., 2014. High performance continuous wave 1.3 μm quantum dot lasers on silicon. Appl. Phys. Lett. 104, 041104.

Liu, A.Y., Herrick, R.W., Ueda, O., Petroff, P.M., Gossard, A.C., Bowers, J.E., 2015. Reliability of InAs/GaAs quantum dot lasers epitaxially grown on silicon. IEEE J. Sel. Top. Quantum Electron. 21, 1900708.

Lourdudoss, S., 2012. Heteroepitaxy and selective area heteroepitaxy for silicon photonics. Curr. Opin. Solid State Mater. Sci. 16, 91.

Luxmoore, I.J., et al., 2013. III–V quantum light source and cavity-QED on silicon. Sci. Rep. 3, 1239–1243.

Mi, Z., Yang, J., Bhattacharya, P., Qin, G., Ma, Z., 2009. High-performance quantum dot lasers and integrated optoelectronics on Si. Proc. IEEE 97, 1239–1249.

Miller, D.A.B., 2000. Rationale and challenges for optical interconnects to electronic chips. Proc. IEEE 88, 728–749.

Monat, C., et al., 2002. nP-based two-dimensional photonic crystal on silicon: in-plane Bloch mode laser. Appl. Phys. Lett. 81, 5102–5104.

Palit, S., Kirch, J., Tsvid, G., Mawst, L., Kuech, T., Jokerst, N., 2009. Low-threshold thin-film III–V lasers bonded to silicon with front and back side defined features. Opt. Lett. 34, 2802–2804.

Schwartz, B., Oehme, M., Kostecki, K., Widmann, D., Gollhofer, M., Koerner, R., Bechler, S., Fischer, I.A., Wendav, T., Kasper, E., Schulze, J., Kittler, M., 2015. Electroluminescence of GeSn/Ge MQW LEDs on Si substrate. Opt. Lett. 40, 3209–3212.

Seassal, C., Rojo-Romeo, P., Letartre, X., Viktorovitch, P., Hollinger, G., Jalaguier, E., Pocas, S., Aspar, B., 2001. InP microdisk lasers on silicon wafer: CW room temperature operation at 1.6 μm. Electron. Lett. 37, 222–223.

Tanabe, K., Guimard, D., Bordel, D., Iwamoto, S., Arakawa, Y., 2010. Electrically pumped 1.3 μm room-temperature InAs/GaAs quantum dot lasers on Si substrates by metal-mediated wafer bonding and layer transfer. Opt. Express 18, 10604.

Tanoto, H., Yoon, S.F., Lew, K.L., Loke, W.K., Dohrman, C., Fitzgerald, E.A., Tang, L.J., 2009. Electroluminescence and structural characteristics of InAs/In0.1Ga0.9As quantum dots grown on graded Si1 − xGex/Si substrate. Appl. Phys. Lett. 95, 141905.

Usman, M., Reithmaier, J.P., Benyoucef, M., 2015. Site-controlled growth of GaAs nanoislands on pre-patterned silicon substrates. Phys. Status Solidi A 212, 443–448.

Wang, T., Liu, H., Lee, A., Pozzi, F., Seeds, A., 2011. 1.3-μm InAs/GaAs quantum-dot lasers monolithically grown on Si substrates. Opt. Express 19, 11381–11386.

Wang, Z., Tian, B., Pantouvaki, M., Guo, W., Absil, P., Van Campenhout, J., Merckling, C., Van Thourhout, D., 2015. Room-temperature InP distributed feedback laser array directly grown on silicon. Nat. Photonics 9, 837.

Wirths, S., Geiger, R., von den Driesch, N., Mussler, G., Stoica, T., Mantl, S., Ikonic, Z., Luysberg, M., Chiussi, S., Hartmann, J.M., Sigg, H., Faist, J., Buca, D., Grützmacher, D., 2015. Lasing in direct bandgap GeSn alloy grown on Si (001). Nat. Photonics 9, 88.

Wu, M., Trampert, A., Al-Zoubi, T., Benyoucef, M., Reithmaier, J.P., 2015. Interface structure and strain state of InAs nano-clusters embedded in silicon. Acta Mater. 90, 133–139.

Yoon, J., et al., 2010. GaAs photovoltaics and optoelectronics using releasable multilayer epitaxial assemblies. Nature 465, 329–333.

CHAPTER THREE

Transfer Printing for Silicon Photonics

Brian Corbett*[,1], Ruggero Loi*, James O'Callaghan*, Gunther Roelkens[†]
*Tyndall National Institute, University College Cork, Cork, Ireland
[†]Ghent University-Imec, Technologiepark-Zwijnaarde, Ghent, Belgium
[1]Corresponding author: e-mail address: brian.corbett@tyndall.ie

Contents

1. INTRODUCTION

While photonics is a key enabling technology for multiple applications, unlike for electronics, the photonic devices for each application are generally customized with the key components being singly packaged. A laser is a vital component in many applications but since, for different applications, the laser emission may be required to be at a wavelength within the UV to the far IR range, this results in a special material structure and the laser generally being a stand-alone component. This singulation of

Semiconductors and Semimetals, Volume 99
ISSN 0080-8784
https://doi.org/10.1016/bs.semsem.2018.08.001

devices requires costly, complex assembly to achieve photonic subsystems. Integration of photonic components is the clear solution. This is not so straightforward due not only to the wide range of materials needed but also to the wide variety of desired photonic functions.

Communications provide an example where it is imperative to reduce costs to allow the continued scaling of information flow. Data flow has been exponentially increasing for decades and while higher costs have been tolerated for high bandwidth internet communications this is not the case as photonics is replacing electrical signaling at shorter distances in fiber-to-the-home, in data centers and ultimately for chip-to-chip communications. The occurrence of the low loss window for silica fiber in the 1300–1550 nm wavelength range requires devices based on InP which provide the appropriate bandgaps for lasers, modulators, and detectors. These devices are the basis for all fiber-based long distance communications. InP wafers have only scaled slowly in size due to the challenges associated with creating the substrates. Currently most production is on 75–100 mm diameter substrates which are costly when compared with GaAs and especially silicon substrates. While this wafer size is adequate for the production of individual lasers and detectors it imposes a severe challenge for circuit integration since photonic device sizes are relatively large and the device arrangements are complex. The optimal device structures are distinct for different photonic functions often requiring epitaxial overgrowth steps and thereby affecting yield. Thus, the number of functional circuits per wafer is low. While excellent progress has been made with photonic integrated circuits (PICs) using InP platforms for telecommunications (Nagarajan et al., 2010); these are expensive to produce with only a handful of foundries producing InP PICs. The scalability of this approach for low-cost photonics is ultimately limited.

The most successful large-scale integration approach for photonics is based on silicon starting from wafers with diameters from 200 to 300 mm using highly mature technological processes that are precise on the nanometer scale. Several of these platforms are available from large, state-of-the-art foundries with established design rules for external designers. Nevertheless, these platforms are not complete without an active gain providing element which is the best supplied from III–V materials. This chapter will discuss the current state of development for microtransfer printing (μTP) as an effective technology for the scalable heterogeneous integration of different materials (particularly III–V semiconductors) and devices onto these silicon platforms. The technique involves the formation and registration of thin coupons of a material structure, releasing these coupons from their source wafer,

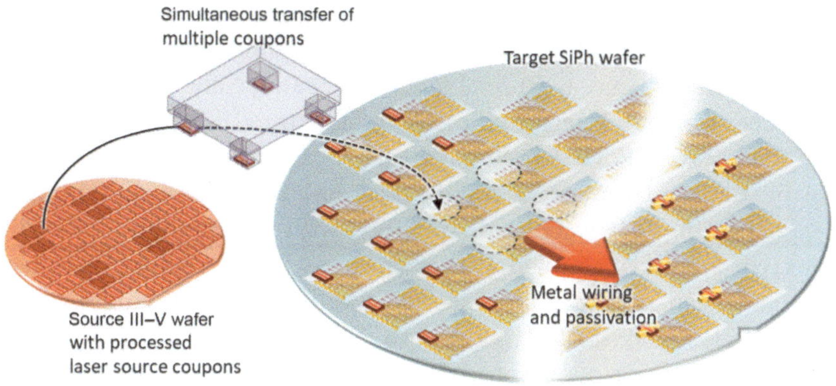

Fig. 1 Heterogeneous integration concept for the transfer of III–V components (materials or devices such as lasers) from a source wafer to a target silicon photonics (SiPh) wafer.

the sequential, parallel transfer of stamp-selected arrays of the coupons, and their bonding to selected matching locations on the target silicon substrate as shown schematically in Fig. 1.

1.1 Silicon Platforms

The silicon-on-insulator (SOI) technology has matured over the last 20 years to be a scalable platform for photonics (Reed and Knights, 2004). It takes advantage of the enormous development in technology for electronics and imposes the learnings on design rules and circuit verification. It relies on a thin layer of silicon surrounded by SiO_2 acting as a waveguide layer. The thickness of the Si is commonly 220 nm though other examples work with silicon layer thickness up to 4000 nm where coupling to external lasers and fibers is more straightforward. The accessible wavelengths are from 1.1 to 4 μm with the principle interest being for datacenters and communications around 1550 and 1300 nm. However, the same waveguide technology can be used for many other applications such as LIDAR, gas sensing, and medical devices (Wang et al., 2017).

The light source or an optical amplification function has yet to be monolithically integrated on silicon on a wafer scale. While recent advances have been demonstrated with selective area epitaxy and quantum dot growth on silicon (Liu et al., 2014), these approaches have yet to be proven in photonic circuits. In the meantime, the laser light has to be provided to the silicon waveguides from a separate InP (or GaAs)-based material structure.

Laser light can be coupled to these silicon waveguides using the following three general techniques. Grating coupling requires alignment and fixation of optical fibers or external lasers. Recently, fully formed VCSELs have been directly integrated on the wafer using flip chip technology (Lu et al., 2016). Edge coupling between a gain chip and the waveguide can be designed for broadband connection. By controlled etching of trenches in the SOI wafer and combined with accurate flip chip of InP reflective amplifier chips high-performance external cavity lasers have been achieved on large cross-section (3 μm) SOI waveguides (Zilkie et al., 2012) and small cross-section SOI waveguides (Guan et al., 2018). Hybrid lasers allow for a more scalable integration. In these lasers, there is an evanescent optical connection between the Si waveguide and a III–V gain element. The integration uses bonding of centimeter-sized dies up to full III–V wafers to the SOI wafer. This bonding approach has been well studied (Roelkens et al., 2010) and successfully commercialized. It does however limit the complexity of the SOI wafer as the surface needs to be flat (planarized) over the large area being bonded. It is also relatively inefficient in the use of costly III–V material especially if the component need is sparse.

More advanced SOI technologies have demonstrated modulators using lateral p–n junction, carrier accumulation/depletion with metal-oxide-semiconductor capacitor structures and the Franz–Keldysh effect in Ge grown on silicon. High bandwidth detectors have been widely used by local deposition of high-quality Ge. Yet these functions could be provided with higher performance by III–V materials if they could be integrated in a sensible manner. So while the gain providing components are uniquely provided by III–V materials there are many opportunities for the integration of other established photonic components based on InP, GaAs, and GaSb materials. It may also be possible to introduce desired novel functions based on optimized nonlinear materials such as $LiNiO_3$ or $BaTiO_3$. Furthermore, it could even be possible to integrate compact electrical circuits using III–V or Si technology for very high bandwidth electrical drivers and amplifiers. The power for optical circuits could be remotely delivered through fibers if an efficient power converter element was integrated on the circuit. Thus, a technology where different components can be heterogeneously cointegrated is highly desirable. This can be delivered through μTP.

The second set of platforms is based on the use of Si as a simple substrate but using well-established materials with wafer scale deposition and etching processes. Waveguide systems based on doped silica waveguides, silicon nitride (SiN), and polymers are well established. These waveguide materials allow accessing a wider range of wavelengths particularly in the visible region. As there is minimal absorption by water in the visible region, this opens up many applications in life sciences especially in combination with microfluidics. Foundry processes for SiN are available. They provide relatively compact circuits with low waveguide loss allowing for larger propagation lengths and larger circuits. Refractive index control—as needed for tunable functions—can be provided by integrated heaters. The gain element is lacking and, as above, multiple active components spanning a wider material range including GaAs and GaN integrated at low cost are desirable.

The ultimate integration is the combination of photonics with the most advanced CMOS electronics. Electrical signal interconnections limit the bandwidth performance of modern electronic processing systems. Higher bandwidth can be achieved if photonics can be closely integrated with the electrical functions such as the processor or memory. However, such integration must not compromise the advanced and highly specialized electrical circuitry. The CMOS chips can be integrated on a silicon platform for waveguide interconnections or the waveguides can be directly integrated with the CMOS. In both cases the challenge of integrating the laser remains.

In summary, to obtain the most versatile, functional optical circuits require the integration of diverse materials and devices (Table 1).

Table 1 Platform Technologies Requiring Light Source Integration

Si Platform	Waveguide Form	Wavelength	Applications
Silicon-on-insulator	220 nm SOI	1100–4000 nm	Data communications
	Thick SOI		Computer interconnect
			Sensors
			Phased arrays
Si substrates	Polymer	300–3000 nm	Sensors
	Doped SiO_2		Biomedical
	SiN_x		Computer interconnect
CMOS electronics	Advanced technology nodes	800–1600 nm	Short reach, high bandwidth, and high density interconnect

2. TRANSFER OF EPILAYERS AND MICROTRANSFER PRINT

The III–V epilayers contain all the functionality required for device operation and so their separation and transfer have been long practiced. Epitaxial lift-off using an ultrathin AlAs release layer is well established for GaAs-based photovoltaics. Photonic recycling in thin structures results in enhancement of the open circuit voltage and record performance for single junction cells has been achieved with this approach (Kayes et al., 2011). Additionally, the cost of the substrate can be recovered and so epitaxial lift-off is now implemented commercially by a number of companies. Nanomembranes created especially for group IV materials is an active field of research (Rogers et al., 2011). Through the release and transfer of very thin structures, it is possible to apply stress profiles to Si materials in a manner that would not otherwise be possible (Scott and Lagally, 2007). These techniques use large areas of material without specific focus on alignment to the target substrate. Bonding of III–V material structures to a target substrate (e.g., SOI) followed by substrate removal and device fabrication is a very successful approach to the integration of active III–V materials with silicon waveguides (Roelkens et al., 2010) and is dealt with in another chapter in this volume. The μTP is an alternative approach to integration and this chapter will present the characteristics of the technique and its advantages.

μTP starts with the formation of arrays of islands of material (termed coupons hereafter) from a source wafer. These coupons, of appropriate dimensions, that can be actual processed devices, are tethered to the substrate, holding them in place while they are released from the source wafer by the selective etching of a buried layer (or the substrate) in the structure. Following this, a sequential, parallel, pickup, and transfer of stamp-selected arrays of the suspended coupons to selected matching locations on a target substrate is performed. The target substrate may be prepatterned but is locally flat for the receiving coupon. The process sequence for the transfer of a single coupon is indicated schematically in Fig. 2.

Note that the technique transfers the coupons in an "epitaxial-side-up" configuration in contrast to a wafer bonding approach. The primary requirement for bonding to the new substrate is that the mating surfaces are locally flat. This can be achieved by spin coating the surface of the target substrate with a thin layer of polymer which can also assist the adhesion of the coupons.

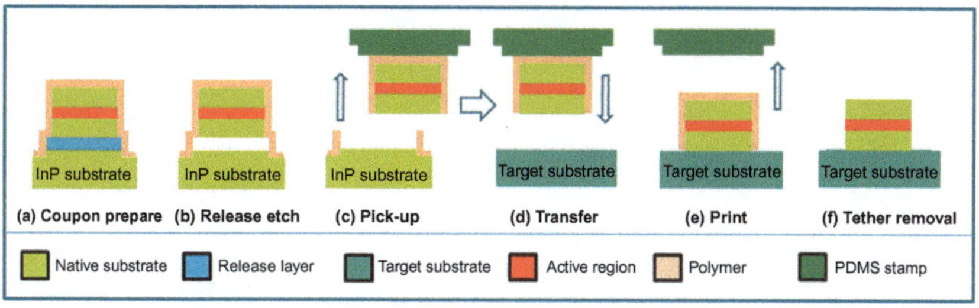

Fig. 2 Schematic of the microtransfer print process indicating the preparation, pickup, and transfer of a single coupon from a source substrate (InP) to a target substrate.

A direct bond can also be achieved between the mating surfaces through van der Waals forces. These are particularly enabled for very flat, thin, and small dimensioned coupons. Transfer print has been demonstrated for multiple materials including silicon, III–V materials and devices (GaAs, InP, InAs, GaN), graphene, and dielectrics. One example of the integration of large numbers of different components on a single substrate is for large area displays. Separate red (GaAs), green, and blue (GaN) micro-LEDs together with a silicon control circuit have been integrated for each pixel in the display (Bower et al., 2017) with thousands of devices being transferred per step. The magnification afforded by the printing of small devices on a large pitch makes possible the very effective use of the source material.

The important features of μTP for silicon photonics are:

- The highest quality source materials, as used for conventional devices, are used. The proviso is that a release layer and associated layers for waveguiding and height control are included. The technique is independent of the diameter of the source substrate and can print to large diameter target substrates.
- It makes efficient use of expensive materials. More devices per unit area can be obtained as the individual devices can be placed within 10–20 μm of each other on the source wafer since space for full bond pads is not required to be on the device.
- The devices can be prefabricated to allow manufacture in an optimum environment and to allow evaluation of their performance prior to transfer and identify known good die. In that case it is probably necessary that bond pads are provided for the device assessment.
- Accurate registration of the coupons on the target wafer is obtained by the use of fiducial markers on the source and target. A positional accuracy of ± 1.5 μm at 3σ can be obtained in high throughput, large area

operation though accuracy better than ± 0.5 μm is achievable in individual coupon placement. Precise (lithographic) alignment to underlying waveguides can be obtained by postprocessing the associated waveguides on the transferred coupon.

- High throughput is obtained through wafer scale transfer of 2D arrays of devices in parallel. The step and repeat cycle takes <40 s.
- Multiple source materials can be selected in sequence. Thus lasers, amplifiers, modulators, detectors, and electronic components can be transferred. These devices can be transferred adjacent to each other thus allowing arrays of complex photonic circuits to be realized on the target wafers.
- The μTP approach allows the material to be transferred to structured substrates such as fully processed target wafers once the target area is locally flat.

2.1 Background to the Microtransfer Technique

A major advance in printing was made at the University of Illinois with the demonstration of kinetically controlled adhesion between an elastomeric stamp, an object, and a substrate (Meitl et al., 2006). Together with a technique where coupons of silicon could be held in place, and therefore registered, by silicon tethers which fractured in a controllable manner (Menard et al., 2004) allowed their transfer in a controllable, massively parallel manner with micron scale accuracy from a source wafer to a separate target substrate.

There are a number of excellent reviews on the technique and its general applications (Carlson et al., 2012; Yoon et al., 2015). The technique utilizes the viscoelastic properties of elastomeric materials and the associated motion-dependent adhesion properties. A master stamp is prepared with the array pitch required and is replicated with polydimethylsiloxane (PDMS) rubber. The stamp contains posts with dimensions corresponding to the desired size of coupons to be transferred. The stamp is retained on a transparent rigid support plate allowing the use of a vision recognition system for accurate transfer of the devices. The stamp is brought in contact with the coupons. A fast pulling motion is used to break the tethers at engineered locations picking up the selected coupons which are now adhered on the stamp posts. The coupons can be transferred and printed with the selected pitch/repetition to the target substrate and the stamp released by a slow movement. The stamp can be utilized repeatedly to consume all the devices/coupons from the source wafer. The coupons can be transferred to multiple different target substrates.

Transfer printing has been demonstrated for many different devices in different configurations. Silicon solar cells with thickness of 10–20 μm and lateral dimensions of 50–100 μm × 1.4 mm have been transferred to glass and other substrates (Yoon et al., 2015). LEDs have been transferred to rigid and flexible substrates (Kim et al., 2011; Trindade et al., 2015). Near record efficiency solar cells have been achieved through stacking (Sheng et al., 2014). A five-junction solar cell comprised of a triple-junction GaAs-based cell mechanically stacked on a dual-junction GaSb/InGaAsSb cell was reported (Lumb et al., 2017) with an implied cell efficiency of 44.5%.

Many works have shown transfer of devices to flexible substrates.

μTP was first commercialized for the production of concentrator photovoltaic cells. Large-scale transfer tools were developed to achieve the production of the large volumes of multijunction cells which were transferred to metal substrates where world record performing solar modules were demonstrated (Furman et al., 2010). X-Celeprint (x-celeprint.com) is developing and licensing patented μTP technology.

2.2 Preparation of Devices for Microtransfer Print to Si Photonics

The μTP technique is implemented for Si photonics by the preparation of arrays of coupons on the source material that are then encapsulated and released from their native substrate by the selective etching of a sacrificial layer while the devices are held in place using a tethering system. Following transfer of the coupons, postprocessing on the target wafer may be required (Fig. 3). The entire process needs to be properly designed at the outset as there are several sequences possible. A key aspect to consider is the final coupling between the device and the waveguide. Butt coupling, evanescent coupling, and grating coupling to devices have all been implemented each with a specific relative positional tolerance requirement. The structuring of the material into devices (etching, contacts, etc.) can be performed in advance of the coupon preparation. This allows for the source wafers to be processed in a specialist foundry on the relevant wafer size without

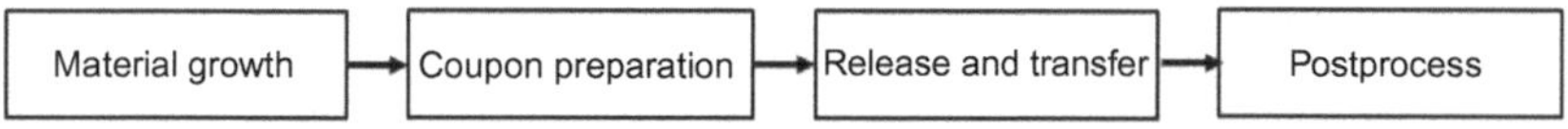

Fig. 3 Overview of process flow for the heterogeneous integration of diverse materials by microtransfer print.

limitations in the process technology and the devices to be optimally manufactured. In addition, this strategy allows for the devices to be tested in part or in full before transfer. This approach relies on the position tolerance of the transfer process and postprocessing steps to connect the devices. If the material alone is transferred then the alignment can be lithographically defined posttransfer. This approach requires foundry compatibility with the process technologies needed for the newly introduced materials and for the processing to take place on the new larger target wafers. The decision as to the best approach may also be determined by the density of transferred coupons.

Other considerations relate to the stress introduced though the process steps which are more significant due to the thin coupons used. They need to be managed in order to achieve the flattest coupons and thus the highest bonding yield and printing accuracy. The dimensions of coupons depend on the device type where, for example, long and narrow coupons are needed for laser or gain or waveguide-coupled elements. The target location is considered in relation to the optical, electrical, thermal, and mechanical interfaces. The devices can range from 20 to 800 μm on a side for grating-coupled photodetector and photovoltaic cells, respectively. The TP technology may not be as advantageous for larger devices.

2.2.1 Materials to be Transferred for Photonic Devices

Epitaxial structures grown on GaAs, InP, GaN, GaSb are used to create devices at relevant wavelengths. In addition, Ge deposited on Si provides a route to high-speed photodetection. Other photonic materials can be deposited on appropriate temporary substrates. In all situations, the μTP process seeks the inclusion of a release layer which could be the source substrate itself. The thickness of the release layer is chosen to be in the range of 500–1000 nm influenced by the ultimate bowing of the coupons after release which in turn is dependent on the stresses introduced by the fabrication processes. The subsequent layers are designed for the final device.

2.2.2 Preparation of Coupons

At a minimum it is necessary to etch mesas to define the coupons and to provide access to the release layer. The sidewalls of the coupons may need to be encapsulated to protect the other layers of the structure from being etched during the release process. If the devices are preprocessed then the various steps of etching waveguides, etching to buried contact layers,

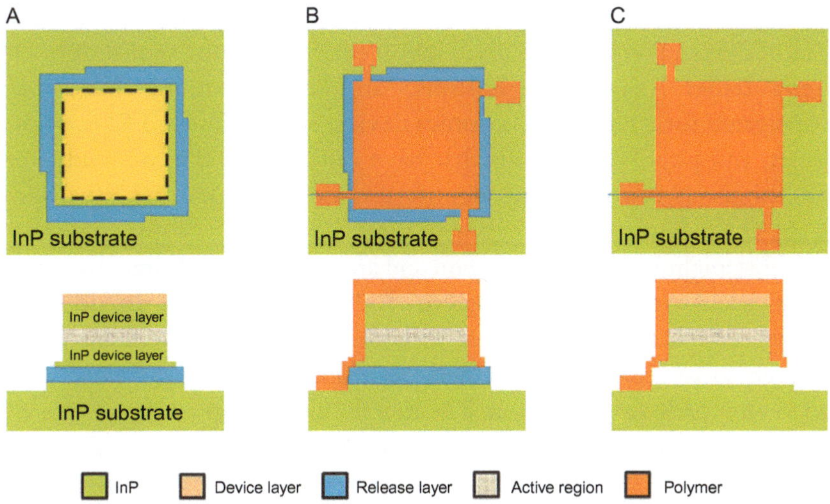

Fig. 4 Example process preparation of InP coupons for transfer (plan and cross-section at shown location). (A) Define respective mesas for the device and the release layer. (B) Resist tethering arrangement anchors the coupons to the substrate and prevents complete sealing of the release layer. (C) The release layer is etched resulting in suspended coupons.

formation of facets, deposition of dielectrics, and contact metallization are performed. Crucial to the eventual success is in managing the stresses in the coupons which can cause them to bend when released. Control can be achieved by deposition of layers with appropriate compensating stress. The overview of an example of the process sequence is shown in Fig. 4.

2.2.3 Tethering Systems

An anchoring system is needed to keep the devices stable and registered in place during their release from the source substrate. The anchor system must provide protection to the coupon, access for the etchant to the sacrificial layer, adequate adhesion to the substrate, stability during the undercut, rinsing, and drying steps and break easily and cleanly when the devices are being picked up. The material for the anchor system can be organic polymers such as standard photoresists that are shaped by lithography and made stiff by soft or hard baking. These materials can accommodate the height differences involved. Other choices for the anchor material are silicon dioxide, SiN, or the semiconductors themselves which can provide higher mechanical stability in a wider range of cases but can be more difficult to implement.

The engineering of a suitable anchor system strongly depends on the dimensions of the coupons, the materials involved, and the final application. Different tether shapes, dimensions, and configurations have been investigated for a range of photonics devices such as lasers, LEDs, photodiodes, and photovoltaic cells that have different thicknesses and lateral dimensions. It is possible to identify few general rules that usually can be applied to most of the anchor systems.

Resist anchors cover the coupon and are held to the substrate by a number of tethers. Some exposed regions on the sacrificial layer provide access for the etchant for the undercut to commence. The tethers are formed between these regions. The tethers are made by two main parts, a foot and a connection to the coupon. The foot should be large enough to provide adhesion to the substrate and not be peeled off during the pickup. Engineering of strong adhesion gradients along the narrowing tethers allows them to break in the desired location which is usually as close as possible to the coupon as this prevents flaps and debris which could affect the printing. The thickness of the polymer depends on the area of the coupons and the stiffness desired. The heat treatment of the resist alters its stress and mechanical properties. Finally, the number of tethers should be adjusted according to the perimeter of the coupon in order to withstand the capillary forces involved during the undercut.

An example of anchors for InP coupon transfer consists of 5 μm wide tethers arranged on a 40 μm pitch. The connection to the foot is kept of the same width for the first 1 μm to allow tolerance against misalignments during contact lithography without creating differences in the tether width, then is tapered down to 2 μm in order to allow a second breakage spot for the tether before encountering the foot which has to be the largest possible. The access regions protrude about 5 μm. The width of the tether foot is limited by this width. The feet can be enlarged to touch each other when 5 μm from the grating, this provides enough space for the etchant to access the sacrificial layer with proper transport properties. Coverage of the substrate and the free spaces between anchors of different coupons is recommended to reduce potential contamination by excess etched material. A third breakage point can be introduced along the foot by creating a slit. Also, V shapes at the meeting point between the connection and the foot can offer a higher gradient in the case of a tilted peeling of the tethers. Fig. 5 shows an example layout while Fig. 6 shows optical images of the tether system applied to the release of InP device coupons.

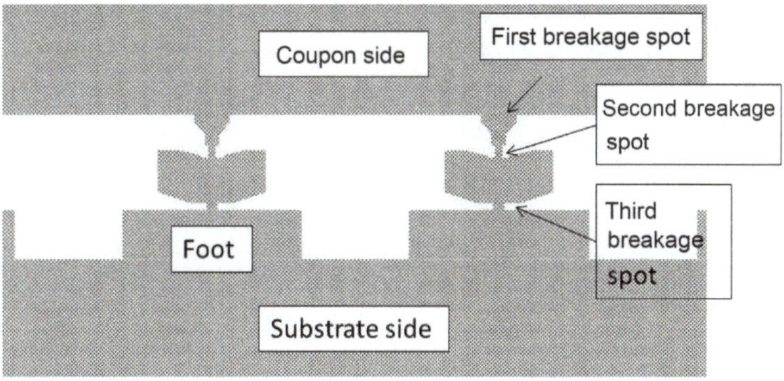

Fig. 5 Layout example of a tether retaining a coupon in place during and after undercut etching.

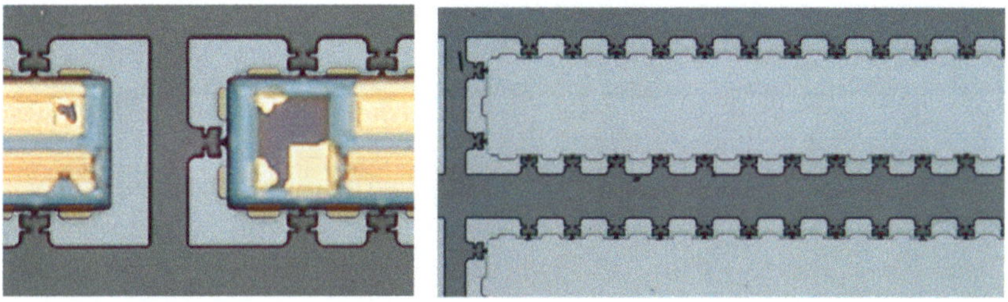

Fig. 6 Microscope image of resist tethering system showing devices held in place after release etch and before pickup (*left*) and after device pickup showing remaining InP stub and resist anchors (*right*).

2.2.4 Release Etching

The coupons can be released using dry or wet etching techniques. Dry techniques include the selective etching of Si and Ge in XeF_2 vapor which is selective against surrounding layers of SiO_2 and GaAs, respectively. More commonly, the coupons are released using wet chemical etching. Very high selectivity and very smooth etching are desired as otherwise the bottom surface of these coupons, i.e., the interface with the target substrate, will become shaped requiring thicker adhesives. Selective wet etchants are known for many material combinations (e.g., Hjort, 1996) and some examples are given in Table 2.

After release etching, the fragile coupons are typically rinsed in water and dried in air. The coupons can collapse during this stage, generally caused by internal stress, and resulting in unpickable coupons so stress management is of critical importance. Remarkably, there is no need for critical point drying

Table 2 Examples of Selective Etchants for Selected Layers Against Their Surrounding Combinations

Layer to be Etched	Surrounding	Etchant
InGaAs	InP	$FeCl_3$
InAlAs	InP	$FeCl_3$
InP	InGaAsP	HCl
InAlP	GaAs	HCl
$Al_xGa_{1-x}As$ ($x > 0.5$)	GaAs	HF, HCl
SiO_2	Si	HF (liquid or vapor)
Si	SiO_2	Tetramethyammonium hydroxide (TMAH), KOH $XeFe_2$
Ge	GaAs	$XeFe_2$

which is probably due to the fact that the selective etchants used leave the adjoining surfaces hydrophobic. It is also noteworthy that coupons of several hundred microns in dimensions can be undercut with submicron thick release layers and be easily picked up to die to the tethering process. After the removal of all coupons, it is potentially possible to recover the substrates for reuse.

We now discuss the development of a release layer for InP-based epitaxial structures which are the mainstay for the highest performing devices in the 1300–2200 nm wavelength range. The lattice-matched heterostructures provide many combinations of layers which can be selectively etched with the interface between -As and -P containing layers being attractive due to a selection of potential etchants. A release layer of InGaAs was shown to be effective for the realization of multilayer air–InP reflector structures (Kusserow et al., 2008). While many acid mixtures provide selective etching the highest selectivity was obtained using $FeCl_3$:H_2O (2:1) at temperatures below ambient. The optimum temperature is in the range of 5–8°C since the etch rate decreases and precipitation occurs as the temperature is lowered below 1°C.

The selectivity between InGaAs and InP is high for the $\langle 010 \rangle$ direction which is at 45 degree to the wafer flat (011). This requires the coupons be orientated along those directions for the shortest release time and thus the highest selectivity. However, ridge waveguides should be oriented along/ perpendicular to the major flat for taper formation in InP using existing wet chemical processes. This results in release times of many hours, consequent dishing (nonflat profiles) of the released surface, and limiting the width of the coupon. A solution is found through the use of InAlAs as a release layer (O'Callaghan et al., 2017). InAlAs etches at a faster rate in the same $FeCl_3:H_2O$ mixtures, and the etch rate has a low dependence on the crystal direction. Fig. 7 shows optical microscope images comparing partly underetched circular mesas that used 500-nm thick InAlAs layers and InGaAs layers, respectively. No crystallographic dependence for the etching of InAlAs could be observed, while there is substantial dependence in the case of InGaAs. The quality of the bottom surfaces of the coupons can be assessed by atomic force microscopy and white light interferometry. It is found that the growth conditions during the switch from an As layer to a P layer play a role in the smoothness of an InAlAs–InP interface and that an interface layer of InGaAs assists in achieving a smoother interface. The best smoothness was obtained with a 50-nm InGaAs thick layer (Fig. 8). Using this release bilayer structure enables the orientation of coupons along any chosen crystal axis without penalty and leads to flatter surfaces after the release of wider coupons.

A similar approach can be used for the release of GaAs structures using a AlGaAs or AlInP release layer.

An interesting, if unwanted, result can arise if the layers to be released are thin and are strained by, for example, a protection dielectric layer such as SiN_x. The effect in question occurs as the etch release comes to it last stage

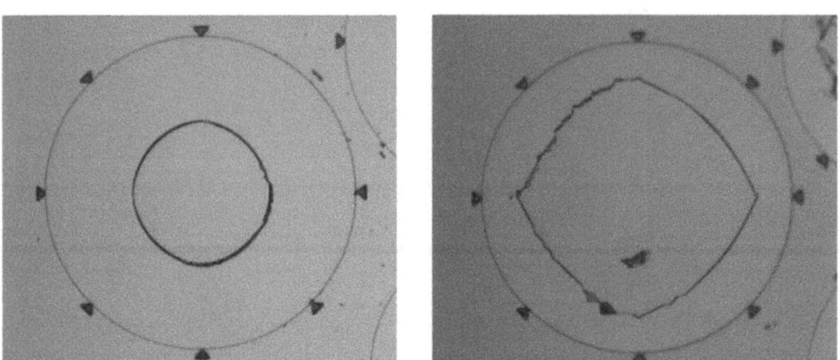

Fig. 7 Microscope images of release process for 300 μm diameter circular mesas at mid-stage following tape-removal of the undercut section. *Left* shows the residual mesa resulting from a 500-nm thick InAlAs release layer where it is evident that the undercut process has only a minor crystallographic dependence and *right*, the residual mesa with a 500-nm thick InGaAs release layer where strong crystallographic dependence is evident. In addition, the etch rate for the InAlAs is many times as that for the InGaAs.

Fig. 8 White light interferogram of the bottom side of an InP coupon after release etching of an InAlAs/InGaAs bilayer.

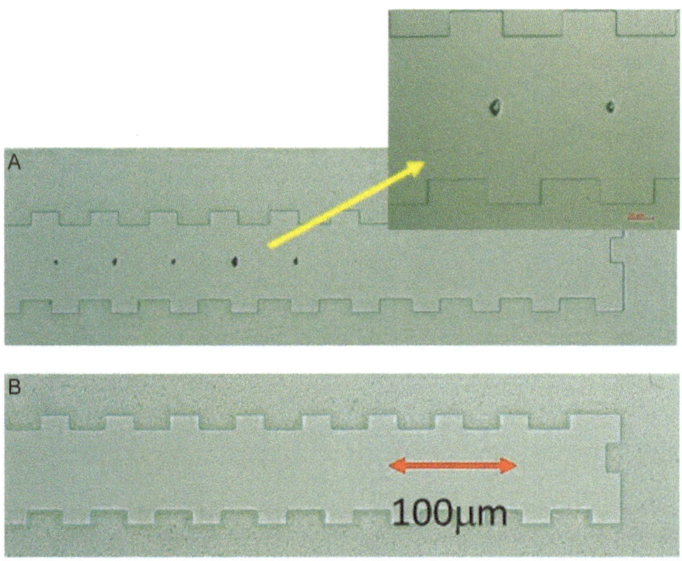

Fig. 9 Optical images of stress induced pillars created after etching with a stressed layer on the coupon (A) and its resolution with a stress managed layer (B).

and pinches off the final part of the release layer. At that stage, there is considerable stress on the coupon at these points which can lead to breaks in the coupon and ingress of the etchant to unwanted locations. In an example of resultant pillars shown in Fig. 9, the upper sample had a thick stressed SiN layer whereas the lower sample had the SiN mask removed and no defects were present. Both samples were etched in $FeCl_3$:H_2O 1:2 for 20 min at 1.5°C.

2.2.5 Transfer Process

A stamp created from PDMS with a single protruding post or an array of protruding posts separated by the required grid arrangement for the target wafer is used to pickup the devices. Note that the source wafer has a higher density of coupons than the target and is structured such that there are an integer number of coupons between those being picked by the stamp. In other words, there is a defined magnification between the source and the target. The grid spacing may be different in the two lateral directions. Picking of the devices is done by pulling back the stamp fast from the source wafer. It is essential that the target location be clean and flat and that the coupons themselves do not have any residue or debris from the tether breaking and undercut etching steps. Overpressure can be applied to the coupons. The coupons are not damaged by this process. The transfer is carried out in air. The coupons are released from the stamp by a slow peel motion. Generally the process is carried out at ambient temperature which is necessary to maintain registration over large areas. Release can be improved at higher temperatures at the expense of alignment and softening of the PDMS above 60°C. The adhesion of the coupons to the target depends on the properties of the mating surfaces. Polymers, including benzocyclobutene (BCB), polyimide, polyurethane, and intervia can be spinned on or spray coated on the target substrate to provide a flat adhesive layer. These polymers have excellent adhesive characteristics with low moisture absorbance, low curing temperature, and low stress. They can also act as a stress buffer. The choice of the adhesion layer thickness is determined by the need for evanescent coupling, the required thermal properties and the mechanical properties needed to accommodate stress. This results in thicknesses for the adhesive between 50 nm and 2 μm. The adhesion can be improved by oxygen plasma or thermal pretreatment and by curing of the polymer posttransfer. Direct, adhesive less, van der Waals adhesion can be achieved if the released surfaces are flat and smooth enough which allows for potential electrical connections and improved heat sinking of the devices.

Alignment accuracy requires high-contrast (metal) registration features on the source and target both within the field of view of the vision system. Positional accuracy of ±0.5 μm has been demonstrated (Ye et al., 2018a). After transfer, the tethers are removed by oxygen plasma in the case of resist. This allows the devices to be probed or to be connected up using standard process steps of planarization, via opening and metal deposition. The transferred coupons will withstand the thermal budget required which is typically <300°C.

3. EXAMPLES OF INTEGRATION

3.1 Etched Facet Laser on Si

Stand-alone lasers that are a few microns thick and that are transferred to a selected substrate are a basic example of the technology. The starting wafer to achieve such a device requires that an etch release layer and a lower cladding of sufficient thickness to isolate the laser mode from the target substrate to be included (Fig. 10). The width of the coupon is set by the etch selectivity. The lasers can be fabricated posttransfer (Justice et al., 2012) or pretransfer (Loi et al., 2016). The standard process steps of defining a ridge waveguide and p- and n-type ohmic contacts need to be supplemented by a means of making a resonator by using etched facets, using resonant or grating structures in the silicon waveguide circuit or within the III–V material. An etched facet was utilized where smooth and vertical facets can be obtained by inductively coupled plasma etching. A high reflection coating on the rear facet can be achieved by the angled evaporation of a reflective metal on a dielectric protection layer. The coupon is made longer to facilitate probing directly on the wafer. Additional steps as discussed earlier are implemented to allow the release of the laser coupons. The lasers are bonded directly to silicon and characterized in continuous wave (CW) mode. Fig. 11 shows the temperature-dependent light–current characteristics of the laser where improved temperature performance is obtained for the transferred lasers compared to the lasers on the source wafer due to the superior thermal conductivity of silicon (130 W/mK) over InP (63 W/mK).

Such laser (or gain) chips can be edge coupled to silicon waveguides as currently done using flip chip on silicon photonic platforms. μTP will allow multiple chips to be placed in close proximity. The laser chips have to be aligned with respect to waveguides where the reference height on the silicon

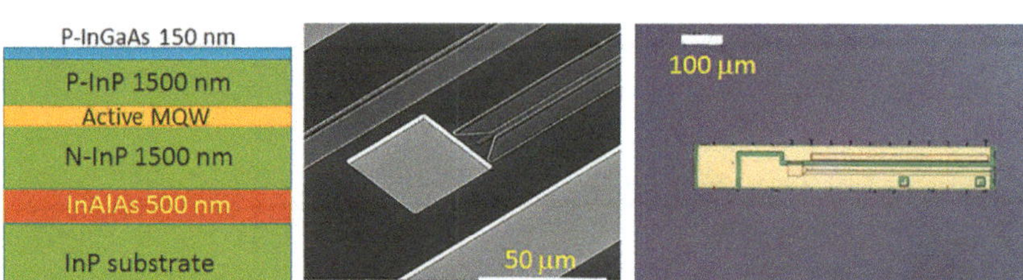

Fig. 10 Schematic layer structure for a 1550 nm emitting laser (*left*), scanning electron microscope image of an etched facet (*center*), and optical microscope image of transferred laser prior to tether removal (*right*).

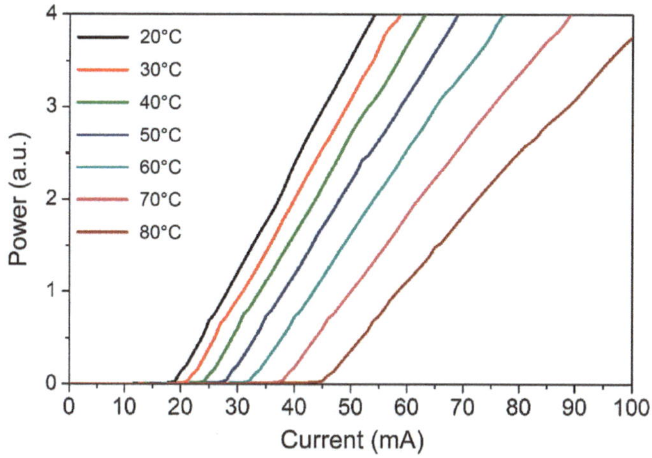

Fig. 11 Temperature-dependent light–current characteristics of 1550 nm etched facet laser transferred to Si.

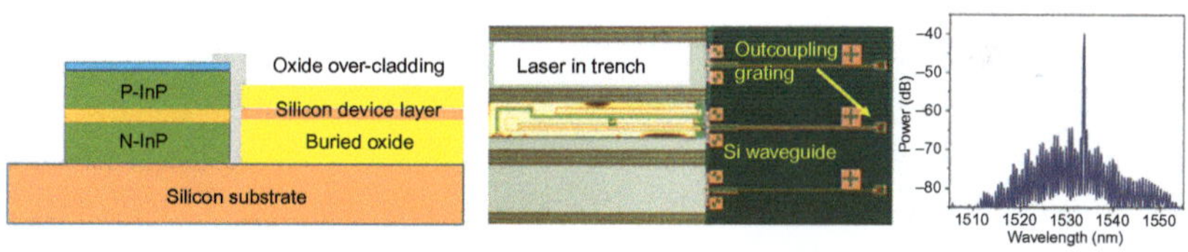

Fig. 12 Edge geometry for laser–waveguide coupling (*left*) using microtransfer printing, optical image of laser printed with respect to the waveguide (*center*), and outcoupled spectrum from grating (*right*).

photonics wafer is controlled by the thickness of the buried oxide layer. The laser waveguide height is then referenced with respect to this surface. The lateral alignment of the laser relies on high-contrast fiducials present on both the laser and the target substrate. Facets are etched in the silicon photonics wafers at the waveguide end using a chrome mask. To obtain the highest coupling efficiency, the laser mode field should be matched to that of the Si waveguide by using spot size converters. Fig. 12 shows an optical microscope image of the aligned laser that is recessed from the waveguide facet by ~3 μm. The spectrum from the laser when outcoupled using the fiber-to-chip grating coupler on the waveguide is shown.

3.2 Evanescent Laser on Si Using a Tapered Coupling

Evanescent coupling between a gain material and silicon waveguide containing a distributed feedback Bragg (DFB) reflector or a ring resonator-based reflector is a second approach for creating an integrated laser. This has been widely implemented with die-to-wafer bonding using tapered waveguides on the silicon and III–V to achieve coupling. The same approach can be implemented with transfer printing noting that the interface to the silicon waveguide is grown on the III–V wafer immediately after the release layer. This allows for the growth of structures for electrically injected components in a p-on-n configuration which is often preferred over an n-on-p configuration required for wafer bonding due to the diffusion of the standard p-dopant (Zn) during wafer growth. The first demonstration of evanescent coupling to SOI waveguides using transfer printing was an optically pumped InGaAsP quantum well structure by De Groote et al. (2016). A thin BCB layer was used to adhere the coupons to the SOI waveguides and allowed for structuring of aligned tapered waveguides for optical coupling. The optically pumped LED showed a 3 dB bandwidth of 130 nm in line with devices realized using a traditional die-to-wafer bonding method. An electrically pumped DFB laser integrated on and coupled to a silicon waveguide circuit using transfer printing was realized by Zhang et al. (2018). Coupons of $40 \times 970\,\mu m^2$ of AlInGaAs gain material were picked from the growth wafer and transferred to SOI waveguides where a second-order grating had been defined with a period of 477 nm and a duty cycle of 75%. Postprocessing was used to define the III–V guiding region, the adiabatic taper structure for coupling to the silicon waveguide layer and the ohmic contacts (see Fig. 13A–C). The resultant laser had a threshold current of 18 mA and a maximum single-sided waveguide-coupled output power above 2 mW. Single mode operation around 1550 nm with >40 dB side mode suppression ratio was demonstrated as shown in Fig. 13D.

An alternative route for the laser is to define the feedback structure in the III–V structure. A 1D nanobeam laser based on a three quantum well InGaAsP gain structure was coupled to a silicon (SOI) waveguide with a coupling efficiency of 83% using transfer printing (Lee et al., 2017). The asymmetric photonic crystal laser was first created by etching nanoholes into the 280-nm thick structure. The 7–8 μm long nanobeams were tethered in place with the thin semiconductor itself during the undercut etching of an InP layer. The nanobeams were then transferred to a structured SOI wafer and placed to bridge a gap in the waveguide. The best positional accuracy was better than 0.3 μm and an angular misalignment of <1.3 degree. The experiment validated simulations that a very short (3.2 μm long) overlap was an efficient vertical coupler. When optically pumped with a 980 nm laser, lasing was obtained at a pump power of 0.2 mW and strong light

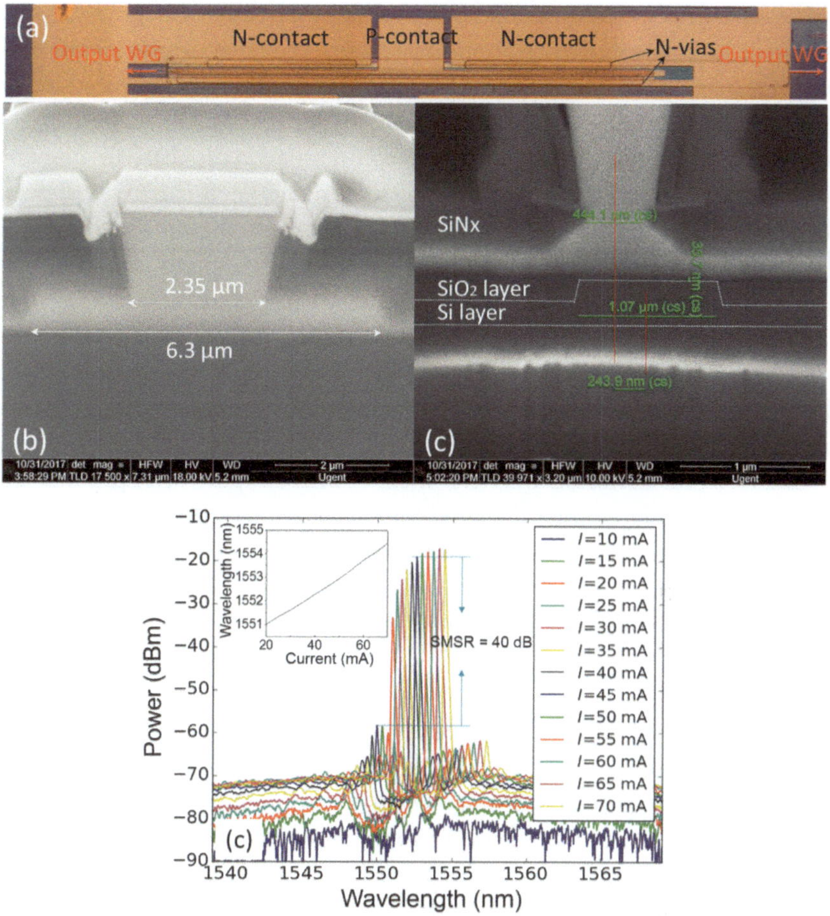

Fig. 13 (A) *Top*-view of the evanescently coupled III–V-on-silicon DFB laser; (B) cross-section of the laser mesa (B) and taper tip (C); (D) output spectrum of the laser as a function of bias current. *Reproduced with permission from Zhang, J., et al., 2018. Transfer-printing-based integration of a III–V-on-silicon distributed feedback laser. Opt. Express 28, 8821–8830.*

emission was measured at the end of the SOI waveguide with a narrow emission peak around 1550 nm.

Photonic crystal nanobeam structures were also fabricated in a 200-nm thick GaAs slab with six layers of InAs quantum dots (Osada et al., 2018). The structure was grown on a 1-μm thick $Al_{0.3}Ga_{0.7}As$ release layer. The nanobeam laser structures were released and transfer printed to SOI waveguides. Lasing emission at 1170 nm was measured under pulsed operation for a low pump power at a temperature of 8 K.

More complex components such as subcircuits can be integrated by evanescent coupling using transfer printing. A first exercise in proving the technique was coupling between a conventional SOI waveguide substrate and coupons containing comparable passive SOI waveguides which looped the light back to the conventional waveguides (Ye et al., 2018a). The coupons were encapsulated in amorphous silicon and used narrow triangular-shaped Si tethers. Stress control was vital to prevent coupon collapse during the HF release etch of the underlying SiO_2. The coupling interfaces were designed to be tolerant of misalignments of ± 1 μm. Adiabatic tapers show a loss per interface of 1.5 dB at 1310 nm while directional couplers showed a loss of only 0.5 dB at 1600 nm. Following that work, the authors demonstrated the integration of waveguide-coupled germanium photodiodes onto passive silicon waveguide circuits (Ye et al., 2018b). In the process Ge p–i–n photodetectors were fabricated on SOI wafers. The coupons (310 μm × 50 μm) contained a trident coupler designed for evanescent coupling with the target SOI. The responsivity of the transferred detector was 0.66 A/W with a bandwidth of 14 GHz. The device was used to show 40 Gb/s transmission. This strategy opens many possibilities for subcircuit integration.

3.3 Grating Coupling Photodiodes on Si/SiN

A third way by which components can be coupled with the Si or SiN waveguides is by using grating coupling. The integration of III–V photodiodes on a silicon-based PIC can be realized by transfer printing substrate illuminated p–i–n or metal–semiconductor–metal (MSM) photodetectors on a grating coupler structure defined in the silicon or SiN waveguide layer. This was recently demonstrated by Chen et al. (2018) where GaAs MSM photodetectors were defined on the GaAs wafer, on top of an $Al_{0.9}Ga_{0.1}As$ release layer. After device encapsulation the release layer is etched using HCl. The released devices were transfer printed to a SiN waveguide circuit containing grating coupler structures to interface with the GaAs MSM device. Microscope pictures of stand-alone surface-illuminated MSM detectors and waveguide-coupled bottom-illuminated devices are shown in Fig. 14. The dark current is 22 nA for a 70 μm × 70 μm device, while it is 7.2 nA for a 30 μm × 30 μm device at 2 V bias. For 850 nm wavelength the responsivity is about 0.1 A/W. 20 GHz bandwidth is measured and open 40 Gbps eye diagrams are realized.

A second example is the coupling light from the SOI to InGaAsP photodiodes with a bandgap absorption of 1370 nm (Zhang et al., 2017).

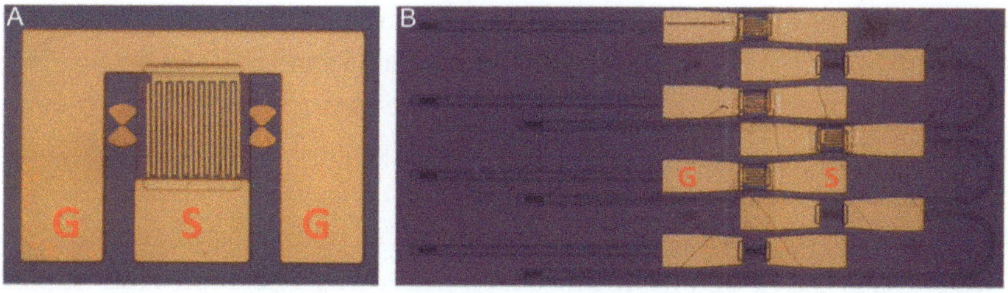

Fig. 14 Panel (A) surface-illuminated 70 μm × 70 μm MSM photodetector; (B) *bottom-illuminated* waveguide-coupled MSM detectors. *Reproduced with permission from Chen, G., Goyvaerts, J., Kumari S., Van Kerrebrouck, J., Muneeb, M., Uvin, S., Yu, Y., Roelkens G., 2018. Integration of high-speed GaAs metal–semiconductor–metal photodetectors by means of transfer printing for 850 nm wavelength photonic interposers, Opt. Express 26, 6351–6359. Copyright 2018 OSA.*

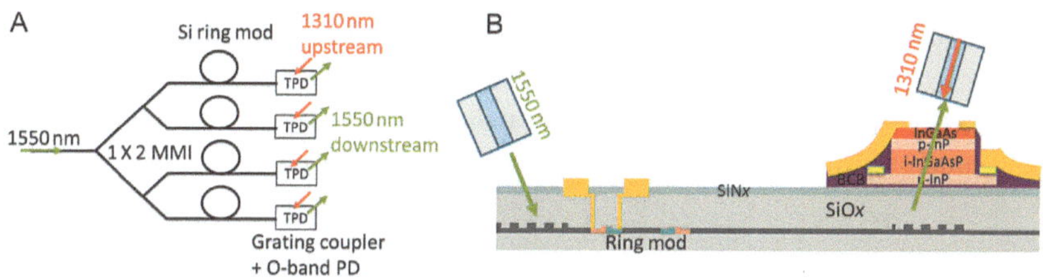

Fig. 15 (A) Schematic layout of the III–V-on-silicon FTTH transceiver array and (B) schematic cross-section of one transceiver. TPD: transparent photodetector. *Reproduced with permission from Zhang, J., et al., 2017. Silicon photonics fiber-to-the-home transceiver array based on transfer-printing based integration of III–V photodetectors. Opt. Express 25, 14290. Copyright 2018 OSA.*

The circuit for a four channel transceiver is shown in Fig. 15 where fiber delivered 1550 nm "downstream" signals are transmitted through the transfer printed 50 μm × 50 μm photodiodes and coupled to a second fiber. That fiber delivers "upstream" signals at 1310 nm. A ring resonator circuit is used to select the downstream wavelengths. The response of the photodiodes at 1310 nm is 0.39–0.49 A/W while at 1550 nm the response is <0.03 mA/W. The 11.5 GHz bandwidth photodiodes were capable of receiving 10 Gb/s signals across the O-band.

3.4 Other Laser Cavity Structures Enabled by Transfer

Microring lasers transferred to glass (Corbett, 1998) and to the end of an optical fiber (Corbett et al., 2002) were demonstrated to lase under CW optical pumping around 1500 nm wavelengths. The microrings with outer diameters of 5.8 μm, width of 1 μm, and a thickness of 0.41 μm contained four quantum wells. The threshold pump power for CW operation was 80 μW. Recently Fan et al. (2017) introduced an intermediate polypropylene carbonate intermediate layer between the devices and the PDMS stamp as a thermal release layer for the transfer of GaAs-based nanoring devices emitting in the 1 μm wavelength range. 100-nm thick AlAs etched in dilute HF served as the release layer for the nanorings. GaAs-based quantum dot-based microring lasers tethered in place by the material itself were transferred to a silver-coated silicon wafer were shown to lase in a plasmonic mode under pulsed operation (Tamada et al., 2017).

A very exciting prospect is the realization of compact VCSEL-like structures. An optically pumped laser emitting around 1450 nm consisting of two high index contrast photonic crystal silicon membrane reflectors with a III–V InGaAsP quantum well heterostructure as the gain medium was realized by transfer printing (Yang et al., 2012). Subsequently, bandedge lasers consisting of a transfer printed active region printed on photonic crystal feedback structures have been realized on both SOI and Si. A threshold of 320 W/cm^2 was achieved for a device emitting at 1540 nm. Such design permits a large mode volume while retaining a single spatial mode.

3.5 Interconnection on Silicon

Polymeric based waveguides are useful for light signaling in datacenters and in high-performance computers at length scales where silicon is too expensive. The light can be coupled to and from silicon waveguides using adiabatic tapered coupling schemes (Dangel et al., 2018). Polymer platforms have also been developed (Zhang et al., 2015) and they need hybrid integration of active components to advance the functionality. μTP can assist with the heterogeneous integration of such components. An example of a simple optical interconnect consisting of an LED and photodiode connected by a polymer waveguide is shown in Fig. 16. LEDs are preferred due to the lack of a threshold current. Wavelength around 1.55 μm is initially selected to avoid any absorption in silicon circuits. The optical interconnect is realized by prefabricating the LED and photodetector and then transfer printing the devices to defined alignment locations on the target silicon substrate. The circuit is designed to operate as a low power stand-off optical switch and with 300 μA in the LED a photocurrent signal of 0.3 μA is obtained in the photodiode.

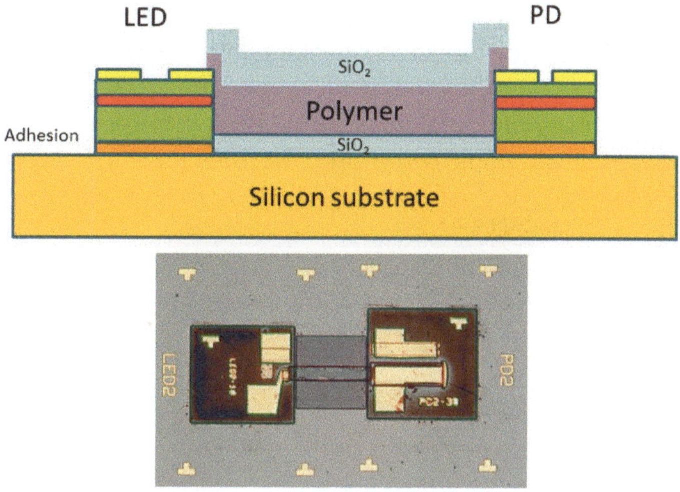

Fig. 16 *Upper* is a schematic cross-section of the optical link containing transfer printed LED and PD while *lower* is the realized connection.

4. CONCLUSIONS

μTP is emerging as a versatile technique for the heterogeneous integration of active and passive components on silicon. It enables the close integration of unique materials not possible by other means. The use of the technique requires engineering of the integration strategy, careful preparation of the starting wafers, and optimization of technological processes for accurate, stress-free transfer of materials, and devices. Alignment is a key issue of silicon photonics and can be addressed through postprocessing of the coupons or by the design of tolerant coupling structures. While many examples of its use for silicon photonics have been shown here it is noted that the work is very recent and is still in an early stage of development. Many more devices can be envisioned to be integrated such as optical isolators, filters, and electronics. The close proximity of the driving electronics can improve signal delivery and reduces the size of bond pads thereby reducing capacitances and allowing higher bandwidth.

Array selection is ideally suited to the production of large numbers of circuits on each wafer thus allow photonic circuits to be produced at low cost for consumer applications. Initial studies have shown that the transferred components pass the challenging reliability specifications. μTP has been used to produce devices in commercial volumes. The technique is now being made available through semiconductor foundries (see microprince.eu).

There are many applications with a need for diverse devices to be cointegrated on a common platform. μTP is an important technology for the development and implementation of new integration concepts.

ACKNOWLEDGMENTS

We acknowledge the valuable contributions from our colleagues at Tyndall and within the TOP-HIT consortium. We especially acknowledge assistance from António-José Trindade and Chris Bower from X-Celeprint. This work is part funded by Science Foundation Ireland under 12/RC/2276 (IPIC) and the European Union's Horizon 2020 Research and Innovation Programme under Grant agreement No. 645314 (TOP-HIT).

REFERENCES

Bower, C.A., et al., 2017. Emissive displays with transfer-printed assemblies of 8μm × 15μm inorganic light-emitting diodes. Photon. Res. 5, 2327. A24.

Carlson, A., Bowen, A.M., Huang, Y., Nuzzo, R.G., Rogers, J.A., 2012. Transfer printing techniques for materials assembly and micro/nanodevice fabrication. Adv. Mater. 24, 5284–5318.

Chen, G., Goyvaerts, J., Kumari, S., Van Kerrebrouck, J., Muneeb, M., Uvin, S., Yu, Y., Roelkens, G., 2018. Integration of high-speed GaAs metal–semiconductor–metal photodetectors by means of transfer printing for 850 nm wavelength photonic interposers. Opt. Express 26, 6351–6359.

Corbett, B., 1998. Spectral characteristics of single-$In_{0.7}Ga_{0.3}As$ quantum-well microring lasers. IEEE Photon. Technol. Lett. 10, 3–5.

Corbett, B., et al., 2002. Lasing from semiconductor microring on the end of an optical fiber. Appl. Phys. Lett. 81, 808–810.

Dangel, R.F., et al., 2018. Polymer waveguides enabling scalable low-loss adiabatic optical coupling for silicon photonics. IEEE J. Sel. Top. Quantum Electron. 24, 8200211.

De Groote, A., Cardile, P., Subramanian, A., Fecioru, A., Bower, C., Delbeke, D., Baets, R., Roelkens, G., 2016. Transfer-printing-based integration of single-mode waveguide-coupled III–V-on-silicon broadband light emitters. Opt. Express 24, 13754–13762.

Fan, F., et al., 2017. Fabrication and room temperature operation of semiconductor nano-ring lasers using a general applicable membrane transfer method. Appl. Phys. Lett. 110, 171105.

Furman, B., et al., 2010. In: A high concentration photovoltaic module utilizing micro-transfer printing and surface mount technology. 36th IEEE PVSC Proceedings, pp. 475–480.

Guan, H., et al., 2018. Widely-tunable, narrow-linewidth III–V/silicon hybrid external-cavity laser for coherent communication. Opt. Express 26, 7920–7933.

Hjort, K., 1996. Sacrificial etching of III–V compounds for micromechanical devices. J. Micromech. Microeng. 6, 370–375.

Justice, J., Bower, C., Meitl, M., Mooney, M.B., Gubbins, M.A., Corbett, B., 2012. Wafer-scale integration of group III–V lasers on silicon using transfer printing of epitaxial layers. Nat. Photonics 6, 610–614.

Kayes, B.M., Nie, H., Twist, R., Spruytte, S.G., Reinhardt, F., Kizilyalli, I.C., Higashi, G.S., 2011. 27.6% conversion efficiency, a new record for single junction solar cells under 1 sun illumination. 37th IEEE Photovoltaic Specialists Conference, pp. 4–8.

Kim, H., Brueckner, E., Song, J., Li, Y., Kim, S., Lu, C., Sulkin, J., Choquette, K., Huang, Y., Nuzzo, R.G., Rogers, J.A., 2011. Unusual strategies for using indium gallium nitride grown on silicon (111) for solid-state lighting. Proc. Natl. Acad. Sci. 108, 10072–10077.

Kusserow, T., et al., 2008. Micromachining of InP/InGaAs multiple membrane/airgap structures for tunable optical devices. Proc. SPIE 6993, 69930B.

Lee, J., Karnadi, I., Kim, J.T., Lee, Y.-H., Kim, M.-K., 2017. Printed nanolaser on silicon. ACS Photon. 4, 2117–2123.

Liu, A.Y., Zhang, C., Norman, J., Snyder, A., Lubyshev, D., Fastenau, J.M., Liu, A.W., Gossard, A.C., Bowers, J.E., 2014. High performance continuous wave 1.3μm quantum dot lasers on silicon. Appl. Phys. Lett. 104, 041104.

Loi, R., et al., 2016. Transfer printing of AlGaInAs/InP etched facet lasers to Si substrates. IEEE Photonics J. 8, 1504810.

Lu, H., et al., 2016. Flip-chip integration of tilted VCSELs onto a silicon photonic integrated circuit. Opt. Express 24, 16258.

Lumb, M.P., et al., 2017. GaSb-based solar cells for full solar spectrum energy harvesting. Adv. Energy Mater. 7, 1700345.

Meitl, M.A., Zhu, Z.T., Kumar, V., Lee, K.J., Feng, X., Huang, Y.Y., Adesida, I., Nuzzo, R.G., Rogers, J.A., 2006. Transfer printing by kinetic control of adhesion to an elastomeric stamp. Nat. Mater. 5, 33–38.

Menard, E., Lee, K., Khang, D.-Y., Nuzzo, R., Rogers, J., 2004. A printable form of silicon for high performance thin film transistors on plastic substrates. Appl. Phys. Lett. 84, 5398–5400.

Nagarajan, R., et al., 2010. InP photonic integrated circuits. IEEE J. Sel. Top. Quantum Electron. 16, 1113–1125.

O'Callaghan, J., et al., 2017. Comparison of InGaAs and InAlAs sacrificial layers for release of InP-based devices. Opt. Mat. Express 7, 4408–4414.

Osada, A., et al., 2018. Transfer-printed quantum-dot nanolasers on a silicon photonic circuit. Appl. Phys. Express 11. https://doi.org/10.7567/APEX.11.072002.

Reed, G.T., Knights, A.P., 2004. Silicon Photonics: An Introduction. John Wiley & Sons, Ltd.

Roelkens, G., Liu, L., Liang, D., Jones, R., Fang, A., Koch, B., Bowers, J., 2010. III–V/silicon photonics for on-chip and intra-chip optical interconnects. Laser & Photonics Rev. 1–29.

Rogers, J.A., Lagally, M.G., Nuzzo, R.G., 2011. Semiconductor nanomembranes: synthesis, assembly, and applications. Nature 477, 45.

Scott, S.A., Lagally, M.G., 2007. Elastically strain sharing nanomembranes: flexible and transferable strained silicon and silicon-germanium alloys. J. Phys. D. Appl. Phys. 40, R75.

Sheng, X., et al., 2014. Printing-based assembly of quadruple-junction four-terminal microscale solar cells and their use in high-efficiency modules. Nat. Mater. 13, 593–598.

Tamada, A., et al., 2017. Demonstration of lasing oscillation in a plasmonic microring resonator containing quantum dots fabricated by transfer printing. Jpn. J. Appl. Phys. 56, 102001.

Trindade, A.J., et al., 2015. Heterogeneous integration of gallium nitride light-emitting diodes on diamond and silica by transfer printing. Opt. Express 23, 9329–9338.

Wang, R., et al., 2017. III–V-on-silicon photonic integrated circuits for spectroscopic sensing in the 2–4 μm wavelength range. Sensors 17, E1788.

Yang, H., et al., 2012. Transfer-printed stacked nanomembrane lasers on silicon. Nat. Photonics 6, 615–620.

Ye, N., et al., 2018a. High-alignment-accuracy transfer printing of passive silicon waveguide structures. Opt. Express 26, 2023–2032.

Ye, N., et al., 2018b. Transfer print integration of waveguide-coupled germanium photodiodes onto passive silicon photonic ICs. J. Lightwave Technol. 36, 1249–1254.
Yoon, J., Lee, S.M., Kang, D., Meitl, M.A., Bower, C.A., Rogers, J., 2015. Heterogeneously integrated optoelectronic devices enabled by micro-transfer printing. Adv. Opt. Mater. 3, 1313–1335.
Zhang, Z., et al., 2015. Hybrid photonic integration on a polymer platform. Photo-Dermatology 2, 1005–1026.
Zhang, J., et al., 2017. Silicon photonics fiber-to-the-home transceiver array based on transfer-printing based integration of III–V photodetectors. Opt. Express 25, 14290.
Zhang, J., et al., 2018. Transfer-printing-based integration of a III–V-on-silicon distributed feedback laser. Opt. Express 28, 8821–8830.
Zilkie, A.J., Seddighian, P., Bijlani, B.J., Qian, W., Lee, D.C., Fathololoumi, S., Fong, J., Shafiiha, R., Feng, D., Luff, B.J., Zheng, X., Cunningham, J.E., Krishnamoorthy, A.V., Asghari, M., 2012. Power-efficient III–V/silicon external cavity DBR lasers. Opt. Express 20, 23456–23462.

FURTHER READING

Fang, A.W., Park, H., Cohen, O., Jones, R., Paniccia, M.J., Bowers, J.E., 2006. Electrically pumped hybrid AlGaInAs-silicon evanescent laser. Opt. Express 14, 9203–9210.
Liu, S.C., et al., 2017. Photonic crystal bandedge membrane lasers on silicon. Appl. Opt. 56, H67–H73.
Liu, L., et al., 2018. On-Chip Optical Interconnect on Silicon by Transfer Printing. CLEO.
Yoon, J., et al., 2008. Ultrathin silicon solar microcells for semitransparent, mechanically flexible and microconcentrator module designs. Nat. Mater. 7, 907.

CHAPTER FOUR

Semiconductor Membrane Lasers and Photodiode on Si

Shigehisa Arai[1], Tomohiro Amemiya[1]
Institute of Innovative Research (IIR), Tokyo Institute of Technology, Tokyo, Japan
[1]Corresponding authors: e-mail address: arai@pe.titech.ac.jp; amemiya.t.ab@m.titech.ac.jp

Contents

1. INTRODUCTION

GaInAsP/InP long-wavelength (1.5–1.6 μm) lasers have been widely used in long-distance optical fiber communications and monolithic integration of these lasers with other functional devices such as electroabsorption modulators and/or monitoring photodiodes (PDs), multiple wavelength laser arrays, and wavelength tunable lasers integrated with microheaters has been widely used in practical systems. Recently, hybrid integration of many functional photonic components including active photonic devices is in progress for cost-effective production of compact photonic integrated circuits (PICs) (Heck et al., 2011, 2013). On the other hand, short-reach (a few meters to a few kilometers) optical fiber communications, which consist of a huge number (several tens of thousands to several

Semiconductors and Semimetals, Volume 99
ISSN 0080-8784
https://doi.org/10.1016/bs.semsem.2018.07.002

hundreds of thousands) of semiconductor lasers, have been used in supercomputers which require huge capacity signal transmission (Doany, 2012; TSUBAME2.5, 2013).

Looking at progress in large-scale integrated circuits (LSIs), the total chip power dissipation became grossly affected by the metal wiring (Banerjee and Mehrotra, 2001; Borkar and Chien, 2011; Ho et al., 2001), and high-speed signal transmissions with an energy cost close to 0.3 pJ/bit were demonstrated in 2008 (Maekawa et al., 2008; Mensink et al., 2007). In order to solve wiring problems of future LSIs, a possibility of optical interconnects was proposed in 1980s (Goodman et al., 1984), but semiconductor lasers with low power consumption operation were not available at that time. After a demonstration of ultra-low threshold current vertical cavity surface-emitting laser (VCSEL) in 1995 (Yang et al., 1995), the possibility of semiconductor lasers for on-chip optical interconnects became more realistic (Chen et al., 2007; Miller, 2000; Ohashi et al., 2009). In 2000, it was predicted that the available energy cost to send one-bit signal would be less than 100 fJ/bit (Miller, 2000), which was much lower than the energy cost of electrical wires at that time (Rattner, 2006). In 2009, device requirements of on-chip optical interconnects became more severe because the energy cost of electrical signal transmission has been improved much (Miller, 2009).

A lot of works were reported on high-speed optical detectors, optical modulators, and light sources with high-speed and low power consumption operation. High-speed operations of Ge (Kang et al., 2009; Rouvìre et al., 2005; Vivien et al., 2009; Yin et al., 2007) and InP/InGaAs (Binetti et al., 2010) photodetectors prepared on SOI substrates were demonstrated. As for modulators based on Si, a Mach-Zehnder interferometric modulator with a 3 dB cut-off frequency higher than 30 GHz (Liao et al., 2007), a four-channel wavelength division multiplexed microring modulator with a speed of 50 Gb/s (Manipatruni et al., 2009), and a complementary metal-oxide semiconductor (CMOS) modulator with an optical pulse energy of 400 fJ/bit under a driving voltage of 1 V (Zheng et al., 2010) were demonstrated. As for light sources prepared on Si or SOI substrates, a Raman silicon laser (Rong et al., 2005, 2007) and a direct gap transition from Ge-on-Si (Liu et al., 2009, 2010) under an optical pumping were demonstrated. Room-temperature continuous-wave (RT-CW) operations of injection-type III–V semiconductor lasers prepared on Si substrates were demonstrated by a wafer direct bonding method (Wada and Kamijoh, 1996), an epitaxial growth (Cerutti et al., 2010; Liebich et al., 2010;

Razeghi et al., 1988; Sugo et al., 1990), benzocyclobutene (BCB) polymer bonding (Roelkens et al., 2006), and low-temperature oxygen plasma-assisted bonding (Fang et al., 2006). Facet-free lasers evanescently coupled with Si waveguides were demonstrated for monolithic integration on SOI wafers (Fang et al., 2007, 2008; Liang et al., 2009). However, since the threshold current of these lasers was similar to that of conventional double-heterostructure (DH) lasers, namely a few tens mA, they are not suitable for a low energy operation of 100 fJ/bit or the less. For this purpose an extremely low threshold current operation as well as high-speed modulation capability is required; hence, semiconductor lasers with very small volume of the active region and high-reflectivity mirrors (high Q cavities), such as VCSELs (Iga, 2000), microdisk lasers (Fujita et al., 2000; Spuesens et al., 2009; Van Campenhout et al., 2007, 2009), and photonic crystal (PC) lasers (Jeong et al., 2013; Matsuo et al., 2010; Park et al., 2004; Takeda et al., 2013), can be possible candidates. An extremely low threshold current even below 10 μA was demonstrated in VCSELs and PC lasers.

We proposed a new kind of semiconductor lasers, namely, membrane distributed feedback (DFB) lasers, which consist of a thin semiconductor core sandwiched by low refractive index cladding layers so as to enhance an optical confinement factor of the active region by a factor of around 3 compared with that of conventional DH lasers, and demonstrated very low threshold operation under an optical pump in 2001 (Okamoto et al., 2002). Various studies of membrane DFB lasers operating under an optical pump have been reported (Naitoh et al., 2007; Okamoto et al., 2003, 2004; Sakamoto et al., 2005, 2007a,b,c). Since very low threshold current operation is essentially required for light sources of on-chip optical interconnects, we introduced a lateral current injection (LCI) structure (Oe et al., 1994) into membrane DFB lasers by direct bonding on an SOI substrate (Maruyama et al., 2006; Okumura et al., 2007, 2009, 2011) and by growth on a semi-insulating InP substrate (Okumura et al., 2010; Shindo et al., 2011a,b, 2013a,b). Even though the latter devices showed threshold current similar to that of conventional DH lasers because the optical confinement factor into the active region is not enhanced, an internal quantum efficiency, η_i, of 70% and a waveguide loss, α_{WG}, of 3.5 cm^{-1} were obtained with Fabry–Pérot cavity lasers (Futami et al., 2012). Relatively low threshold current operation (0.23 mA) was obtained with a membrane DFB laser bonded with BCB on an SOI substrate, but its differential quantum efficiency for an output from the front side, η_{df}, was poor (5%) (Inoue et al., 2015).

Then, it was increased by introducing a distributed reflector (DR) structure consisting of a DFB section integrated with high-reflectivity distributed Bragg reflector (DBR) section (Hiratani et al., 2015b).

In this chapter, theoretical and experimental characteristics of LCI-type membrane DFB and DR lasers intended for ultra-low power consumption optical communication systems such as on-chip optical interconnections are presented. Fig. 1 shows a conceptual diagram of the membrane photonic integrated circuit on Si-LSI. First, fundamentals of the membrane laser such as an enhancement of the optical confinement factor of the active region and its effect on threshold current and modulation speed are discussed in Section 2. Then energy cost analysis of the membrane DR laser is discussed in Section 3. Fabrication and static and dynamic lasing characteristics of membrane lasers are given in Section 4. Sections 5 and 6 are devoted for membrane PDs and an integration of the membrane DFB laser with the p-i-n PD, respectively.

2. SEMICONDUCTOR MEMBRANE LASERS FOR LOW POWER OPERATION

Fig. 2 indicates conceptual cross-sectional structures and fundamental mode profiles of a conventional semiconductor laser and a membrane laser. Since the optical mode field of the membrane laser is compressed into thinner region by a high index-contrast ($\Delta n \sim 40\%$) waveguide structure, an optical confinement factor of the active region, ξ, of the membrane laser is approximately three times higher than that of conventional structure. This enhancement allows to increase the modal gain in the membrane laser and

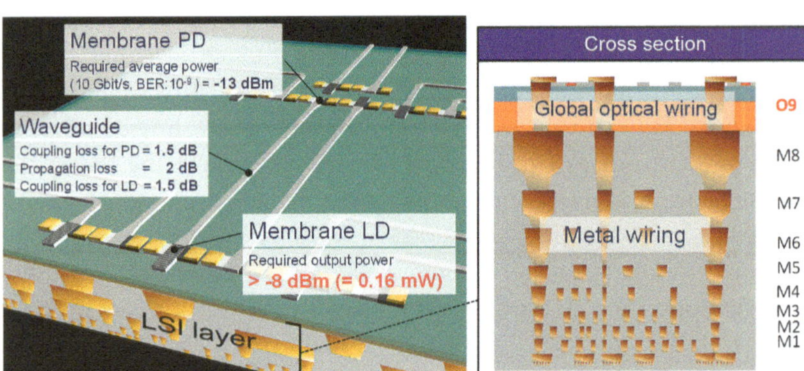

Fig. 1 Image of membrane photonic integrated circuit (MPIC) on a Si-LSI.

will lead to higher modal gain at a fixed injection carrier density or to low threshold current operation when a short cavity structure is adopted. Not only by the enhancement of the modal gain but also by an enhancement of an index-coupling coefficient of the grating in DFB and DBR structures, the active section length of DFB and DR lasers can be shortened to approximately 1/10 times of conventional ones with the same grating depth.

Fig. 3 shows the calculated optical confinement factor of the active region of a slab waveguide in various quantum wells (QWs) as a function

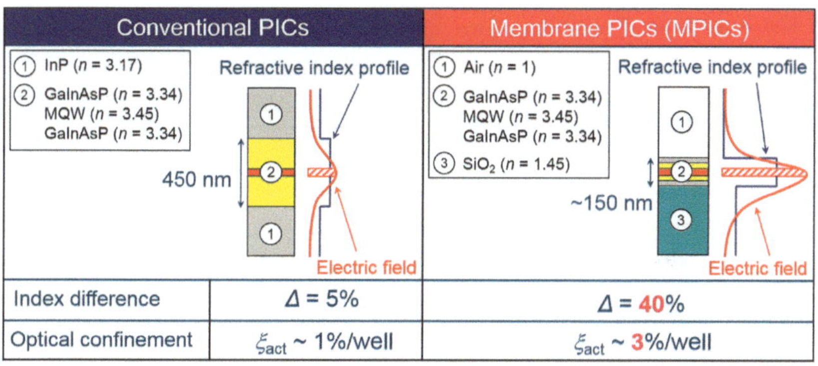

Fig. 2 Comparison of the optical confinement between conventional double-heterostructure laser and membrane laser.

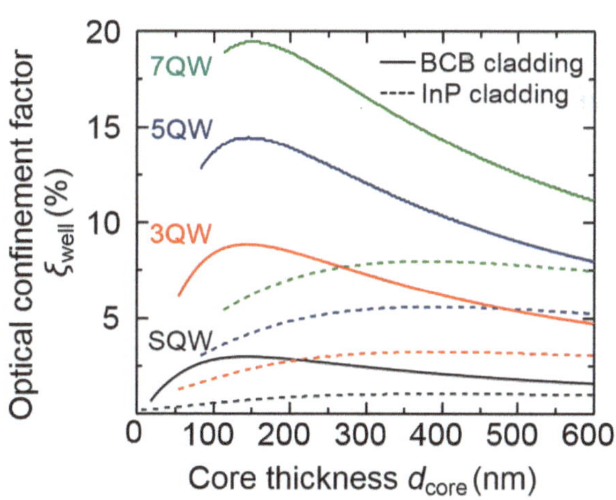

Fig. 3 Optical confinement factor of membrane lasers sandwiched by BCB (*solid lines*) as a function of the semiconductor core thickness. *Dashed lines* indicate that for conventional double-heterostructure sandwiched by InP cladding layers.

of the thickness of the semiconductor core, d_{core} (Okumura et al., 2011). Here, each QW is assumed to consist of 6-nm-thick GaInAsP ($n=3.54$) and 10-nm-thick GaInAsP ($n=3.34$) barriers. As indicated by the solid lines, ξ reaches a maximum value of around 3%/well at d_{core} of around 150 nm and gradually reduces with an increase of d_{core}, whereas that of conventional lasers increases with d_{core} and saturates at around 1%/well.

The enhancement of the optical confinement factor of the membrane laser is very effective not only for reduction of threshold current but also for an increase of direct modulation speed. The relaxation oscillation frequency, f_{r}, of directly modulated semiconductor laser can be expressed by the following relation:

$$f_{\mathrm{r}}=\frac{1}{2\pi}\sqrt{\frac{\xi G'\eta_i}{eV_{\mathrm{a}}}(I_{\mathrm{b}}-I_{\mathrm{th}})}=\frac{1}{2\pi}\sqrt{\frac{\xi(dg/dN)v_{\mathrm{g}}\eta_i}{eV_{\mathrm{a}}}(I_{\mathrm{b}}-I_{\mathrm{th}})}, \tag{1}$$

where G' is the differential gain per unit time, g is the medium gain coefficient, N is the injected carrier density, v_{g} is the group velocity, η_i is the internal quantum efficiency, e is the electronic charge unit, V_{a} is the volume of the active region, and I_{b} and I_{th} are the bias current and the threshold current, respectively. The modulation current efficiency (MCEF) can be given by

$$\mathrm{MCEF}=\frac{f_{\mathrm{r}}}{\sqrt{(I_{\mathrm{b}}-I_{\mathrm{th}})}}=\frac{1}{2\pi}\sqrt{\frac{\xi(dg/dN)v_{\mathrm{g}}\eta_i}{eV_{\mathrm{a}}}}. \tag{2}$$

For a low energy cost operation of semiconductor lasers, the increase of MCEF by the increase of ξ and the reduction of V_{a} is required.

3. ENERGY COST ANALYSIS OF MEMBRANE DR LASERS

In this section, some important results for an energy cost analysis of the membrane DR laser are shown (Hiratani et al., 2015a). By using DFB section length dependences of a threshold current, I_{th}, a differential quantum efficiency for the front side output, η_{df}, the required bias current was calculated for the ultra-low power consumption operation with a high-speed direct modulation of the LCI-membrane DR laser. As mentioned before, a light output power of 0.16 mW at a modulation speed of 10 Gb/s is required by the light source of the on-chip optical interconnection.

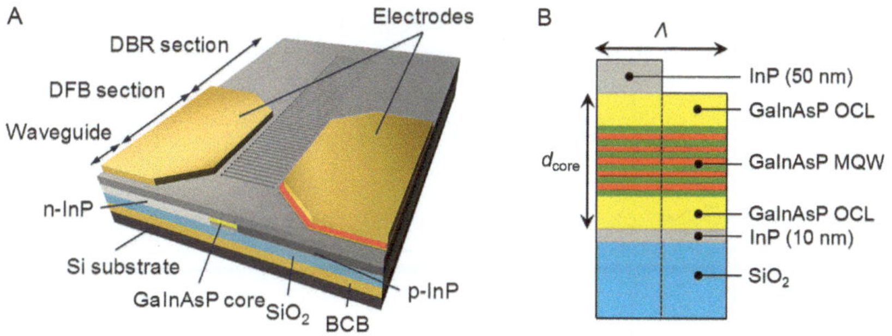

Fig. 4 Schematic structure of (A) a membrane DR laser used for energy cost analysis and (B) its cross-sectional structure of the DFB section.

We assumed that a 3 dB bandwidth $f_{3\mathrm{dB}}$ of 7.7 GHz is required for a bit rate of 10 Gb/s, as the bit rate is 1.3 times higher than the $f_{3\mathrm{dB}}$ in the nonreturn-to-zero (NRZ) signals.

Fig. 4A and B shows a schematic structure of the LCI-membrane DR laser and its cross-sectional structure, respectively, used for the energy cost analysis, where the active region consists of strain-compensated 5-quantum-well (5QW) and a surface grating with a $\lambda/4$ phase-shift region set at the position of $0.3L_{\mathrm{DFB}}$ (L_{DFB} is the length of the DFB section) from the boundary between the DFB and DBR sections because the lower threshold current as well as higher external differential quantum efficiency can be obtained by this $\lambda/4$ phase-shift region.

By varying the index-coupling coefficient, κ_{i}, of the grating in the DFB section and its length, required input power, P_{in}, to the DR laser was calculated for satisfying the condition of the modulation speed of 10 Gb/s ($f_{3\mathrm{dB}}$ of 7.7 GHz) as indicated by open squares in Fig. 5. Here the distance between the active region and p-electrode and the doping concentration of p-InP side cladding layer were assumed to be 1.2 μm and $N_{\mathrm{A}} = 4 \times 10^{18}\,\mathrm{cm}^{-3}$, respectively, and the sheet resistance of it was assumed to be $0.035\,\Omega\,\mathrm{cm}$ for the membrane thickness of 150 nm. Whereas a shorter DFB section (i.e., smaller active region volume) with higher κ leads to lower threshold current and the total power consumption for high-speed modulation can be monotonically reduced as indicated by squares, the total power consumption for the light output power of 0.16 mW becomes much higher for $L_{\mathrm{DFB}} < 50$ μm and it reaches a minimum at a certain length of L_{DFB} because an increase of the power consumption in the device resistance (Joule heating power) becomes dominant in very short DFB section length as indicated by painted circles. Therefore, the total power consumption satisfying the both conditions can be higher value for a certain value of L_{DFB}.

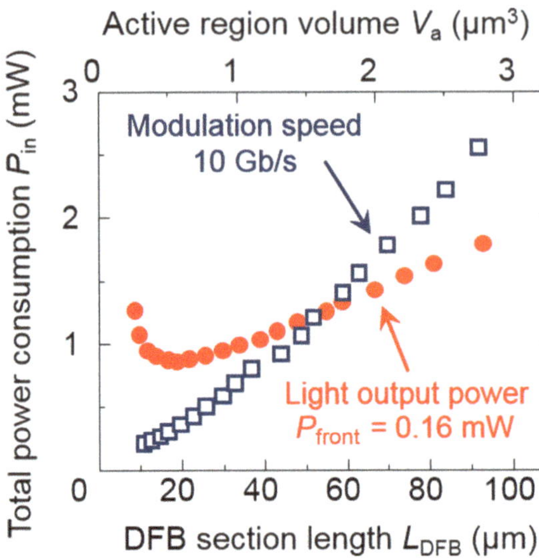

Fig. 5 Total power consumption for the required light output power and that for 10 Gb/s operation as a function of the DFB section length. *From Hiratani, T., Shindo, T., Kyohei Doi, Atsuji, Y., Inoue, D., Amemiya, T., Nishiyama, N., Arai, S., 2015. Energy cost analysis of membrane distributed-reflector lasers for on-chip optical interconnects. IEEE J. Sel. Top. Quantum Electron. 21 (6), 299–308.*

An energy cost [J/bit] of the DR laser can be given by the total power consumption [W] divided by the bit rate [bit/s]. Fig. 6 shows the energy cost of the DR laser for 10 Gb/s signal transmission (indicated by painted circles) and the index-coupling coefficient of the DFB grating (indicated by open squares) as a function of the DFB section length. An energy cost of less than 100 fJ/bit is expected for L_{DFB} of less than 40 μm and a minimum energy cost of 63 fJ/bit can be obtained for $L_{DFB} = 12$ μm, where the active region volume of 0.4 μm^3 and κ_i of 2700 cm^{-1}. For a DFB section length of less than 12 μm, although a lower threshold current is obtained with high κ_i, the total power consumption increases due to the Joule heating.

For further reduction of the total power consumption, a reduction of the optical link loss between the light source and the p-i-n PD and a reduction of the minimum receivable optical power of the p-i-n PD will be required. Fig. 7 indicates the link loss dependence of the energy cost of the DR laser for 10 Gb/s signal transmission and required κ_i of the DFB section. Since the required output power of the DR laser can be reduced by the reduction of the link loss, a structure of shorter L_{DFB} (smaller volume of the active region) with higher κ_i can be adopted. When the link loss is reduced from 5 dB to negligibly small, the energy cost can be reduced from 63 to 28 fJ/bit.

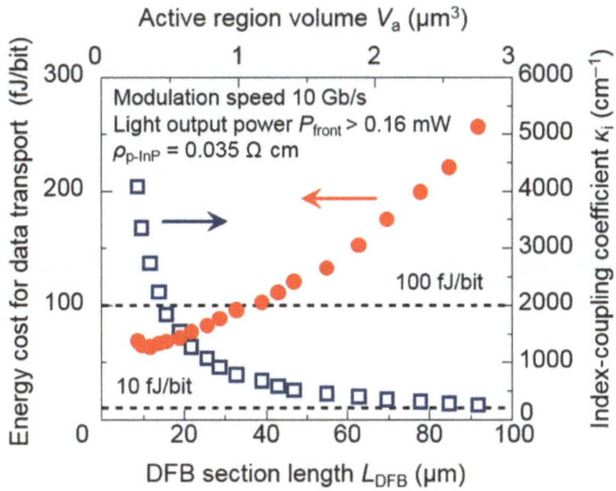

Fig. 6 Energy cost for 10 Gb/s data transport and index-coupling coefficient of an LCI-membrane DR laser as a function of the DFB section length. *From Hiratani, T., Shindo, T., Kyohei Doi, Atsuji, Y., Inoue, D., Amemiya, T., Nishiyama, N., Arai, S., 2015. Energy cost analysis of membrane distributed-reflector lasers for on-chip optical interconnects. IEEE J. Sel. Top. Quantum Electron. 21 (6), 299–308.*

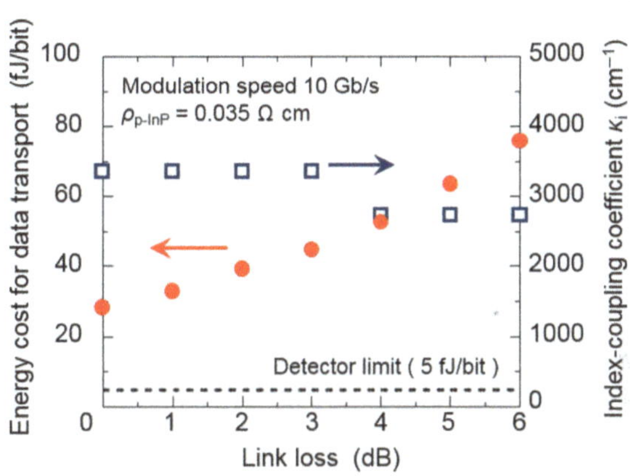

Fig. 7 Minimum energy cost for 10 Gb/s operation and index-coupling coefficient as a function of a link loss. *From Hiratani, T., Shindo, T., Kyohei Doi, Atsuji, Y., Inoue, D., Amemiya, T., Nishiyama, N., Arai, S., 2015. Energy cost analysis of membrane distributed-reflector lasers for on-chip optical interconnects. IEEE J. Sel. Top. Quantum Electron. 21 (6), 299–308.*

4. MEMBRANE DFB AND DR LASERS ON Si

In this section, after a brief explanation of the fabrication process of membrane DFB and DR lasers bonded on a Si substrate, their typical lasing properties such as static lasing properties and direct modulation properties are explained (Inoue et al., 2015).

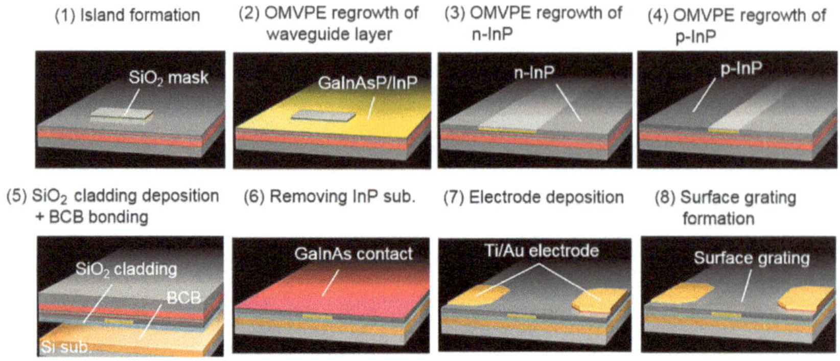

Fig. 8 Fabrication process of a waveguide-integrated LCI-membrane DFB laser. *From Inoue, D., Lee, J., Hiratani, T., Atsuji, Y., Amemiya, T., Nishiyama, N., Arai, S., 2015. Sub-milliampere threshold operation of butt-jointed built-in membrane DFB laser bonded on Si substrate. Opt. Express 23 (6), 7771–7778.*

4.1 Fabrication Process

Fig. 8 shows the fabrication processes of the waveguide-integrated LCI-membrane DFB laser: (1) formation of an island-shaped mesa structure along the [011] direction by CH_4/H_2 reactive-ion etching and selective wet chemical etching, (2) regrowth of GaInAsP waveguide layer for butt-jointed-built-in (BJB) structure by organometallic vapor-phase epitaxy (OMVPE), (3) 7-μm-wide mesa stripe formation and regrowth of an n-InP layer ($N_D = 4 \times 10^{18}\,cm^{-3}$) by OMVPE. After etching one side of the n-InP layer beside the stripe, (4) a p-InP layer ($N_A = 4 \times 10^{18}\,cm^{-3}$) was regrown to form an LCI buried heterostructure with a stripe width of 1–2 μm. Then a 1-μm-thick SiO_2 was deposited onto the wafer by plasma-enhanced chemical vapor deposition followed by (5) the wafer bonding upside-down onto a Si substrate with an intermediate BCB adhesive layer. The BCB layer was prepared by spin coating, followed by curing at 210°C in a N_2 atmosphere. Then, the wafer was bonded to the Si substrate at a pressure of 25 kPa at 130°C. After that, the bonded wafers were baked at 250°C for 1 h in a N_2 atmosphere for hard curing of the BCB adhesive layer. The upper InP substrate side and etch-stop layers were removed by chemical polishing and selective wet chemical etching, thereby revealing the GaInAs contact layer (Fig. 8) (6). After this etching process, the total thickness of the membrane core layers became 270 nm. This contact layer was etched by selective wet etching, except for the p-side electrode region, and the p-InP cap layer on the n-side electrode region was removed to reveal the regrown n-InP layer. Then, (7) Ti/Au electrodes were deposited for both n- and p-type contacts.

Finally, (8) a surface grating structure was formed on the InP cap layer by wet chemical etching with a SiO_2 mask pattern defined by electron-beam lithography. Then, the processed wafer was cleaved into bar form at the waveguide section to measure the light output characteristics. Because the facet had no antireflection coating, the residual reflectivity at the facet is estimated to be approximately 20%.

Fig. 9A shows a scanning electron microscope (SEM) image of the top view of the fabricated waveguide-integrated LCI-membrane DFB laser. Fig. 9B and C shows a schematic diagram of the cross section parallel to the stripe at the coupling section and its SEM image, respectively. Fig. 9D shows an SEM image of the cross section perpendicular to the stripe observed at the passive waveguide section. Although the fabrication process includes a three-step regrowth process, a flat membrane LCI buried heterostructure was successfully obtained.

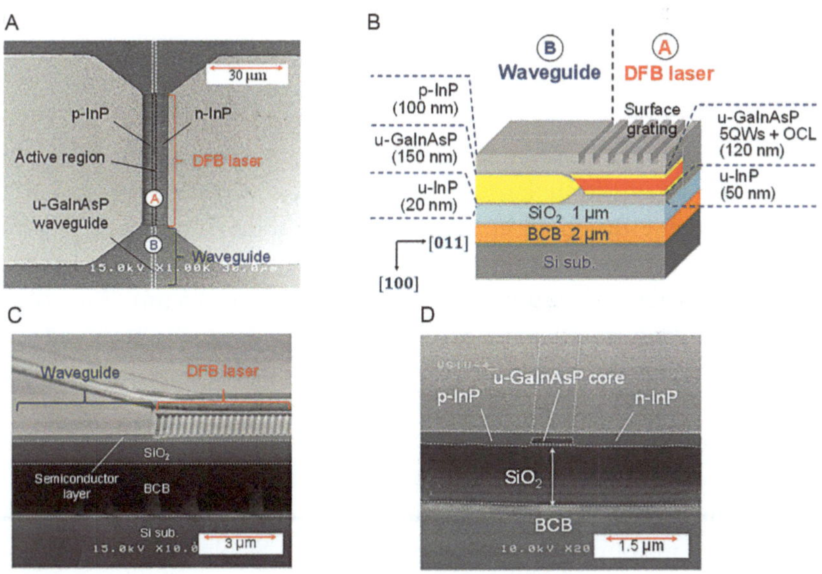

Fig. 9 (A) SEM image of the top view of the fabricated waveguide-integrated LCI-membrane DFB laser. (B) Schematic diagram of the cross section parallel to the stripe at the coupling section. (C) SEM image of the cross section parallel to the stripe observed at the coupling section. (D) SEM image of the cross section perpendicular to the stripe observed at the passive waveguide section. *From Inoue, D., Lee, J., Hiratani, T., Atsuji, Y., Amemiya, T., Nishiyama, N., Arai, S., 2015. Sub-milliampere threshold operation of butt-jointed built-in membrane DFB laser bonded on Si substrate. Opt. Express 23 (6), 7771–7778.*

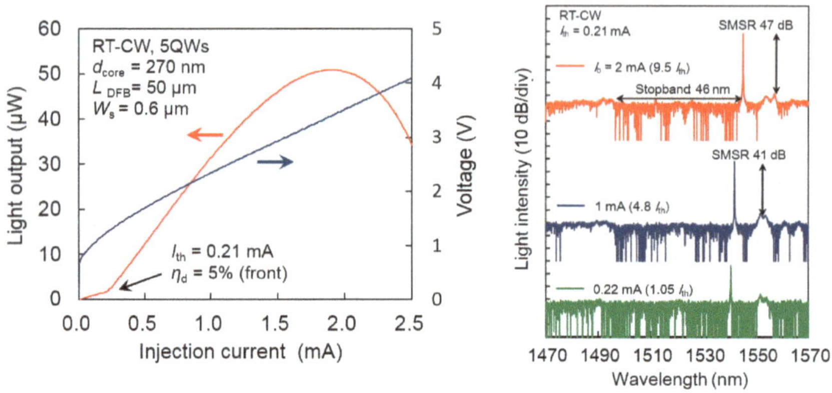

Fig. 10 (A) Light output and voltage–current characteristics of a waveguide-integrated LCI-membrane DFB laser with the DFB section length of 50 μm and the stripe width of 0.6 μm and (B) its lasing spectra at various injection currents. *From Inoue, D., Hiratani, T., Fukuda, K., Tomiyasu, T., Amemiya, T., Nishiyama, N., Arai, S., 2016. Low-bias current 10 Gbit/s direct modulation of GaInAsP/InP membrane DFB laser on silicon. Opt. Express 24 (16), 18571–18579.*

4.2 Static Lasing Properties

Static lasing characteristics of the fabricated membrane DFB laser were measured under RT-CW condition. Fig. 10A shows the light output and the voltage vs injection current characteristics. A threshold current of 0.21 mA and the threshold current density of 700 A/cm^2 (normalized by the number of quantum wells is 140 A/cm^2/well) were obtained with the DFB cavity length of 50 μm and the stripe width of 0.6 μm. An external differential quantum efficiency η_{df} for an output from the front side was only 5%, which may be attributed to the facet phase in respect with the grating period because device had an etched facet without antireflection coating. The operating voltage was reduced to approximately half of that previously reported for the same device length; this can be attributed to the introduction of the alloyed Au/Zn/Au electrode. Fig. 10B shows lasing spectra measured at different bias currents. Stable single-mode operation was observed for a bias current up to 2 mA, after which the light output showed a thermal roll-off. The lasing wavelength was 1542 nm at 1-mA bias current. The side-mode suppression ratio (SMSR) was 41 and 47 dB at bias currents of 1 and 2 mA, respectively. From the lasing spectrum at bias current 2 mA, a stopband width could be determined by measuring the wavelength span where a level of the spontaneous emission intensity was much lower than the noise level of the spectrum analyzer. Based on

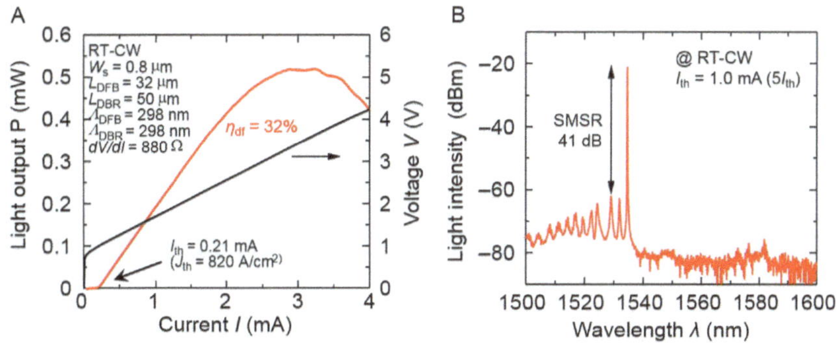

Fig. 11 (A) Light output and voltage–current characteristics of a waveguide-integrated LCI-membrane DR laser with the DFB section length of 50 μm and the stripe width of 0.8 μm and (B) its lasing spectrum at five times the threshold current. *From Tomiyasu, T., Inoue, D., Hiratani, T., Fukuda, K., Nakamura, N., Uryu, T., Amemiya, T., Nishiyama, N., Arai, S., 2018. 20-Gbit/s direct modulation of GaInAsP/InP membrane distributed-reflector laser with energy cost of less than 100 fJ/bit. Appl. Phys. Express 11 (1), 12704.*

the theory of emission spectrum of the DFB laser, the obtained stopband width corresponded to an index-coupling coefficient, κ_i, of 2100 cm^{-1}.

In order to increase the external differential quantum efficiency for the front side output, a membrane DR laser consisting of a DFB section with a DBR at its rear was realized (Tomiyasu et al., 2017). Fig. 11A depicts the current–light and current–voltage characteristics of the device shown in Fig. 4A, under RT-CW operation. The periods of both the DFB and DBR gratings were set to 298 nm, to match the Bragg wavelength in both regions. A threshold current, I_{th}, of 0.21 mA, corresponding to a threshold current density, J_{th}, of 820 A/cm^2, and an external differential quantum efficiency, η_{df}, of 32% for the front side output were obtained. Moreover, the light output from the front side attained a value of 0.16 mW at a bias current of only 0.84 mA. From these curves, a maximum power conversion efficiency of 12.5% was obtained at a bias current of 0.64 mA, and was 12.2% at the bias current of 0.84 mA. The lasing spectrum of the device at a bias current of 1.0 mA (almost five times I_{th}) is shown in Fig. 11B; single-mode operation at 1535 nm with an SMSR of 41 dB was observed. The Bragg wavelength of the DFB section was intentionally not detuned from its gain peak wavelength and the amount of wavelength detuning was estimated to be only 5 nm. The resonant modes in the longer wavelength side of the stopband were not clearly observed because they were approximately 40 nm away from the Bragg wavelength and their optical gains were considerably low. The index-coupling coefficient, κ_i, is estimated to be approximately $1800\,cm^{-1}$, from the stopband width of the other devices fabricated at the same time.

4.3 Direct Modulation Properties

The direct modulation characteristics of the DR laser, whose static properties are shown in Fig. 11, were measured based on small-signal modulation response and 20-Gb/s large-signal modulation (Tomiyasu et al., 2018). First, the small-signal response was measured using a vector network analyzer. An electrical modulation signal was applied to the device via a 40-GHz high-speed ground-signal RF probe with a 100-μm pitch. The optical output signal was coupled to a spherical-lensed single-mode fiber. Finally, the collected optical signal was detected by a 20-GHz p-i-n-TIA (transimpedance amplifier) photoreceiver. Since the relaxation oscillation frequency, f_r, measured by relative intensity noise (RIN) has coincided with that by small-signal response measurement, S_{21} for bias currents ranging from 0.46 to 1.21 mA was measured as shown in Fig. 12A. As the bias current was increased up to 1.21 mA, both the relaxation oscillation frequency, f_r, and 3-dB bandwidth, f_{3dB}, increased. At a bias current of 0.77 mA, f_{3dB} reached 11.2 GHz, which implies that not only 10-Gb/s operation but also 15-Gb/s operation can be possible with a bias current of

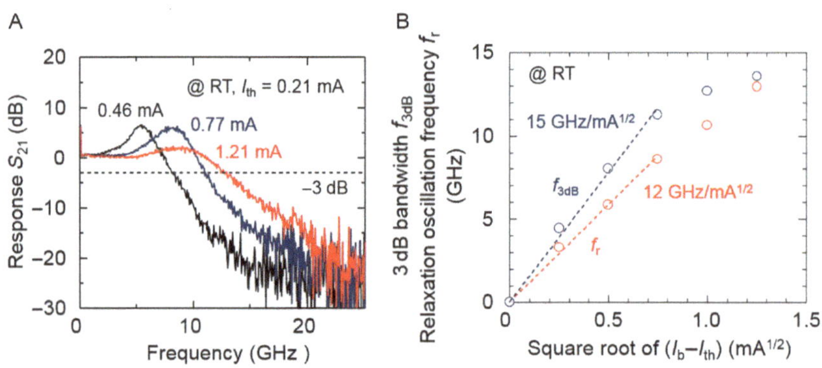

Fig. 12 Small-signal modulation response of the membrane DR laser shown in Fig. 11: (A) responses of S_{21} measured for various bias currents and (B) plots of relaxation oscillation frequency f_r and 3-dB bandwidth f_{3dB} as a function of square root of bias current above threshold. *From Tomiyasu, T., Inoue, D., Hiratani, T., Fukuda, K., Nakamura, N., Uryu, T., Amemiya, T., Nishiyama, N., Arai, S., 2018. 20-Gbit/s direct modulation of GaInAsP/InP membrane distributed-reflector laser with energy cost of less than 100 fJ/bit. Appl. Phys. Express 11 (1), 12704.*

approximately 1.0 mA. The maximum f_{3dB} of 13.6 GHz was obtained at a bias current of 1.77 mA. From the small-signal modulation response, f_r and f_{3dB} were obtained as a function of the square root of bias current above the threshold current, as shown in Fig. 12B. The slopes of fr (modulation current efficiency factor: MCEF) and f_{3dB} were 12 and 15 GHz/mA$^{1/2}$, respectively.

Finally, 20-Gb/s large-signal direct modulation and bit-error-rate (BER) measurements were performed by using an electrical data signal of the NRZ modulation scheme with a $2^{11}-1$ bit word-length pseudo-random bit sequence generated by an arbitrary waveform generator. The optical output was amplified by an Er-doped fiber amplifier (EDFA), and the amplified spontaneous emission was filtered by a tunable bandpass filter. The amplified optical power was monitored and attenuated by using a variable optical attenuator before the 20-GHz p-i-n-TIA photoreceiver. Fig. 13A shows the BER characteristics of the membrane DR laser at a data rate of 20 Gb/s as a function of average received power. The bias current for the device was set to 1.06 mA. The minimum BER of 6.4×10^{-10} was obtained at an average received power of −4.8 dBm. This result demonstrates the feasibility of the membrane DR laser as a light source with ultra-low power consumption for on-chip optical interconnections. Fig. 13B shows an electrical input

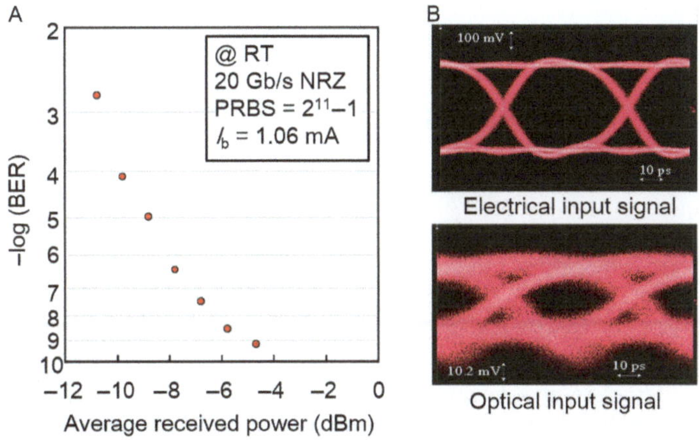

Fig. 13 (A) Measured bit-error-rate (BER) dependence on average received power at 20 Gb/s large-signal modulation and (B) eye diagrams of input electrical signal and optical output measured at a BER of 6.4×10^{-10}. *From Tomiyasu, T., Inoue, D., Hiratani, T., Fukuda, K., Nakamura, N., Uryu, T., Amemiya, T., Nishiyama, N., Arai, S., 2018. 20-Gbit/s direct modulation of GaInAsP/InP membrane distributed-reflector laser with energy cost of less than 100 fJ/bit. Appl. Phys. Express 11 (1), 12704.*

signal and the 20-Gb/s eye diagram measured at an average received power of −4.8 dBm. From a bias voltage of 1.76 V at an injection current of 1.06 mA, the energy cost for this modulation was obtained as 93 fJ/bit; to the best of our knowledge, this is the lowest value reported for membrane DFB and DR lasers. A low energy cost of 97 fJ/bit under 25.8 Gb/s modulation was demonstrated for a similar DR laser bonded on SiO_2/Si substrate, which had a differential resistance of less than 72 Ω and was biased at only 1.04 V (Fujii et al., 2018). If the differential resistance and the bias voltage can be reduced similarly, the energy cost of the DR laser can be further reduced to around 55 fJ/bit (@ 20 Gb/s).

5. MEMBRANE PD ON Si

In this section, after a brief explanation of the fabrication process of membrane DFB and DR lasers bonded on a Si substrate, their typical lasing properties such as static lasing properties and direct modulation properties are explained (Gu et al., 2018).

Fig. 14A shows the schematic structure of the GaInAs/InP p-i-n waveguide-type membrane-based PD bonded on a Si substrate. The device

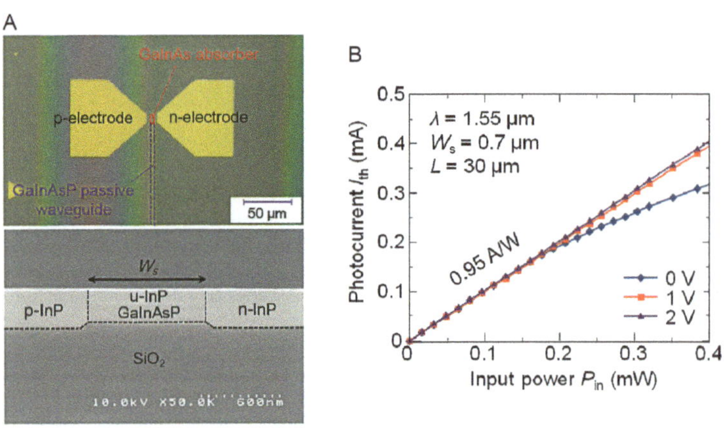

Fig. 14 (A) Structure of a membrane p-i-n photodiode (PD) consisting of absorber region (120-nm-thick GaInAs) integrated with waveguide region (GaInAsP). The stripe width and the absorption region length were 0.7 and 30 μm, respectively. (B) Photocurrent characteristics against input optical power for various applied voltages. *From Gu, Z., Inoue, D., Amemiya, T., Nishiyama, N., Arai, S., 2018. 20 Gbps operation of membrane-based GaInAs/InP waveguide-type p–i–n photodiode bonded on Si substrate. Appl. Phys. Express 11 (2), 22102.*

consisted of a 30-μm-long absorption section with a 120-nm-thick GaInAs bulk layer connected to a passive waveguide section with 155-nm-thick GaInAsP ($\lambda_g = 1.22\,\mu m$) using a BJB structure, and the stripe was 0.7 μm in both sections. Fig. 14B shows the observed photocurrent as a function of input optical power for various applied voltages, where 1.55 μm wavelength TE-polarized light was coupled from a lensed fiber to the membrane PD through the GaInAsP waveguide section. Because the facet of input end was as cleaved without antireflection coating, the coupling efficiency between the lensed fiber to the membrane PD was calculated to be approximately 20%; according to the finite difference method, the responsivity was estimated to be 0.95 A/W.

Fig. 15A shows the BER characteristics of the membrane p-i-n PD for data rates of 10 and 20 Gb/s as a function of average received power, P_{rec}, where NRZ PRBS ($2^7 - 1$) signal and the applied voltage of −3 V were used. The BER of less than 1×10^{-9} was obtained with $P_{rec} = -10$ and −13 dBm for 20 and 10 Gb/s, respectively. As can be seen in Fig. 15B, eye opening was slightly degraded for 20 Gb/s. Much shorter device with smaller capacitance will be required not only for faster operation but also for higher load resistance to increase output voltage.

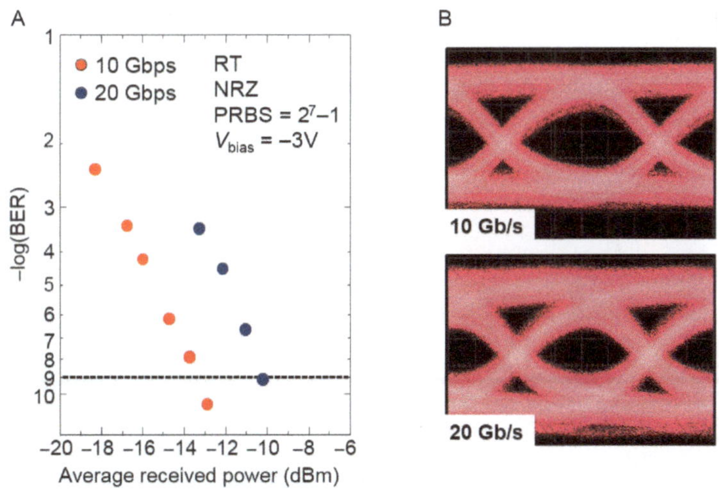

Fig. 15 (A) Measured bit-error-rate (BER) dependences on average received power and (B) eye diagrams for 10 and 20 Gb/s at bias voltage of −3 V. *From Gu, Z., Inoue, D., Amemiya, T., Nishiyama, N., Arai, S., 2018. 20 Gbps operation of membrane-based GaInAs/InP waveguide-type p–i–n photodiode bonded on Si substrate. Appl. Phys. Express 11 (2), 22102.*

6. INTEGRATED MEMBRANE DFB LASER AND p-i-n PD

In this section, a preliminary optical link consisting of a membrane-based DFB laser integrated with a p-i-n PD prepared on a Si substrate is explained (Inoue et al., 2015). The fabrication process is almost the same as that shown in Fig. 8. Fig. 16A shows an optical microscopic image of the fabricated membrane optical link and cross sections at the active and passive regions. The DFB laser and the p-i-n PD had lengths of 80 and 200 μm, respectively. The absorption layer of the p-i-n PD was the same as that of the active region of the DFB laser. These devices were connected by a 500-μm-long passive waveguide. The coupling efficiency between the active and passive section was calculated to be 98%. The propagation loss of the passive waveguide was measured to be more than 8 dB/cm by the Fabry–Perot resonance method using the waveguide with both cleaved facets. The propagation loss for a 500-μm-long waveguide was at least 0.4 dB. The output of the DFB laser on the opposite side to the p-i-n PD was cleaved, for measuring the lasing characteristics. Fig. 16B shows a scanning electron microscopic image of the DFB laser region. As can be seen, the surface grating pattern was successfully formed on the InP cap layer.

The optical link was then statically characterized, by measuring the optical transmission properties between the DFB laser and the p-i-n PD. Fig. 17A shows the light output of the DFB laser measured by an external

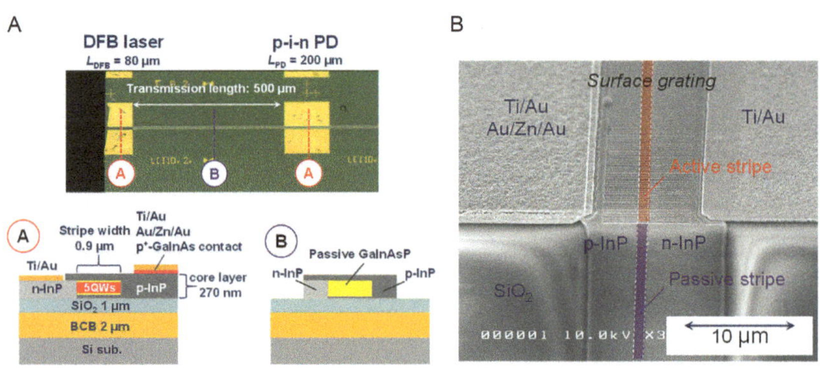

Fig. 16 (A) Optical microscopic image and cross sections of an optical link using a membrane DFB laser and a membrane p-i-n PD. (B) Scanning electron microscopic image of the joint region of the DFB laser and the passive waveguide. *From Inoue, D., Hiratani, T., Fukuda, K., Tomiyasu, T., Gu, Z., Amemiya, T., Nishiyama, N., Arai, S., 2017. Integrated optical link on Si substrate using membrane distributed-feedback laser and p-i-n photodiode. IEEE J. Sel. Top. Quantum Electron. 23 (6), 3700208.*

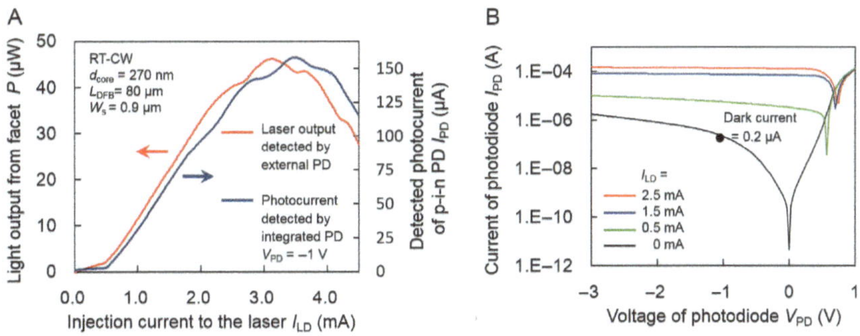

Fig. 17 (A) Photocurrent of the integrated p-i-n PD and light output power of the laser (measured by an external PD) as a function injection current to the laser. (B) Current–voltage characteristics of the integrated p-i-n PD, for various injection currents in the DFB laser. *From Inoue, D., Hiratani, T., Fukuda, K., Tomiyasu, T., Gu, Z., Amemiya, T., Nishiyama, N., Arai, S., 2017. Integrated optical link on Si substrate using membrane distributed-feedback laser and p-i-n photodiode. IEEE J. Sel. Top. Quantum Electron. 23 (6), 3700208.*

PD (on the left) from the opposite side of the p-i-n PD and measured photocurrent, I_{PD}, of the integrated p-i-n PD (on the right), as a function of the injection current, I_{LD}, into the membrane DFB laser. As can be seen, the threshold current of the DFB laser observed from the photocurrent of the integrated p-i-n PD was the same as that measured by the external PD. The slope of I_{LD}–I_{PD} characteristics was 68 µA/mA. Fig. 17B shows the current–voltage characteristics of the integrated p-i-n PD for various injection currents to the membrane DFB laser. The dark current of the p-i-n PD was 0.2 µA at a bias voltage of −1 V, which corresponds to the dark current density normalized by the absorption area of 1×10^{-1} A/cm^2.

Fig. 18A shows the small-signal frequency response (40 MHz to 20 GHz) of the optical link for various injection currents into the membrane DFB laser, with a fixed bias voltage, V_{PD}, of −3 V to the p-i-n PD. A clear relaxation oscillation behavior was observed. In addition, the peak frequency increased with the increase in the DFB laser bias current. Therefore, these responses are not electrical cross-talk between probes, but indeed the transmitted optical signal. The 3-dB bandwidth of the optical link was 11.3 GHz at I_{LD} = 2.73 mA. Fig. 18B shows the small-signal response for various bias voltages applied to the p-i-n PD, for I_{LD} = 2.5 mA. An increase of the reverse bias voltage enhanced the 3-dB bandwidth which can be attributed to carrier extraction by the applied electric field.

Data transmission via the membrane optical link was performed by large-signal direct modulation of the DFB laser. The input electrical signal

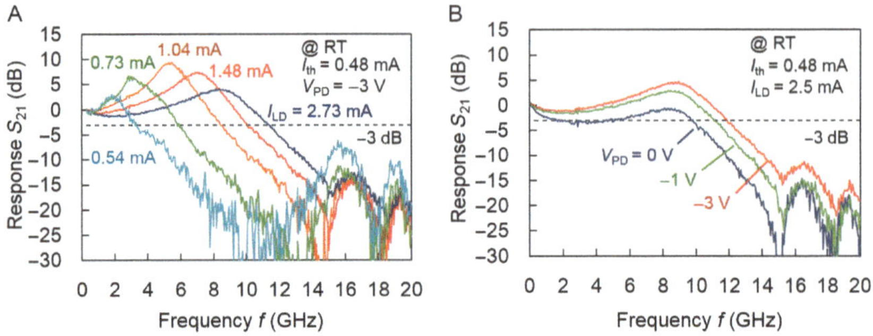

Fig. 18 Small-signal responses of the fabricated optical link; (A) injection current dependence measured at a −3 V bias voltage to the p-i-n PD and (B) bias voltage dependence measured at 2.5 mA injection current of the DFB laser. *From Inoue, D., Hiratani, T., Fukuda, K., Tomiyasu, T., Gu, Z., Amemiya, T., Nishiyama, N., Arai, S., 2017. Integrated optical link on Si substrate using membrane distributed-feedback laser and p-i-n photodiode. IEEE J. Sel. Top. Quantum Electron. 23 (6), 3700208.*

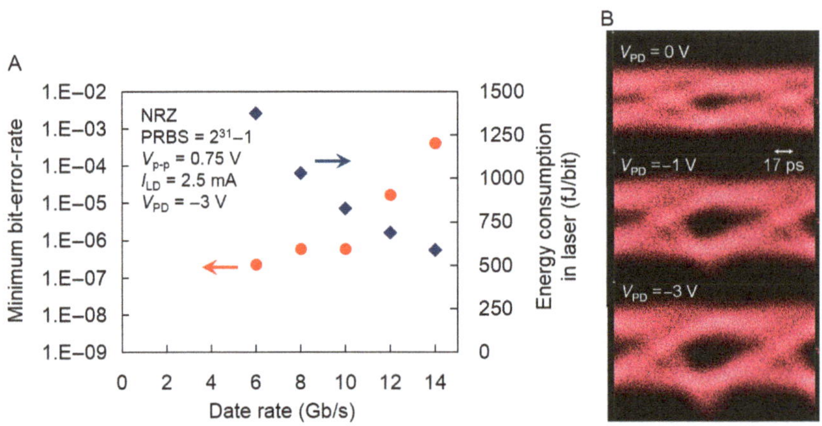

Fig. 19 (A) Bit-error-rate characteristics and energy consumption in the laser of NRZ signal transmission through the membrane optical link, with the laser bias current of 2.5 mA and the p-i-n PD bias voltage of −3 V. (B) 10 Gb/s eye diagrams for the p-i-n PD bias voltages of 0 and −3 V. *From Inoue, D., Hiratani, T., Fukuda, K., Tomiyasu, T., Gu, Z., Amemiya, T., Nishiyama, N., Arai, S., 2017. Integrated optical link on Si substrate using membrane distributed-feedback laser and p-i-n photodiode. IEEE J. Sel. Top. Quantum Electron. 23 (6), 3700208.*

to the DFB laser was NRZ PRBS of $2^{31}-1$ signal, with a voltage swing of 0.75 V_{pp}. Fig. 19A shows the measured BER vs the data rate and the energy consumption in the laser at each data rate. All plots were obtained under the same operating conditions (except for the data rate). A BER in the order of 10^{-7} was obtained up to a data rate of 10 Gb/s. As can be seen in Fig. 19B, clear eye opening was not obtained for $V_{PD}=0$ V and it was much improved for $V_{PD}=-3$ V, which is consistent with the result of the small-signal measurements.

Since the optical link demonstrated here consisted of membrane DFB laser and membrane p-i-n PD with the same active and absorber regions, much higher performance can be expected when the membrane DR laser and the membrane p-i-n PD explained in Sections 4 and 5 are combined.

7. SUMMARY

Our recent investigations of semiconductor membrane lasers and PDs intended for on-chip optical interconnection are explained. A high index-contrast thin semiconductor structure, named semiconductor membrane, allows us to reduce the threshold current of membrane DFB and DR lasers without sacrificing the differential quantum efficiency and to enhance the MCEF to more than two times that of conventional semiconductor lasers. As the result, high-speed direct modulation with the injection current of around 1 mA allows us to demonstrate 20 Gb/s modulation with the energy cost of less than 100 fJ/bit. While there is a room for improvement of the energy cost by reductions of the optical link loss (waveguide loss) and the electrical resistance, it can be reduced to around 30 fJ/bit. For further reduction, the minimum receivable optical power of the membrane p-i-n PD will be a key issue.

REFERENCES

Banerjee, K., Mehrotra, A., 2001. Global (interconnect) warming. IEEE Circuits Devices Mag. 17 (5), 16–32.

Binetti, P.R.A., Leijtens, X.J.M., De Vries, T., Oei, Y.S., Di Cioccio, L., Fedeli, J.M., Lagahe, C., Van Campenhout, J., Van Thourhout, D., Van Veldhoven, P.J., Nötzel, R., Smit, M.K., 2010. InP/InGaAs photodetector on SOI photonic circuitry. IEEE Photonics J. 2 (3), 299–305.

Borkar, S., Chien, A.A., 2011. The future of microprocessors. Commun. ACM 54 (5), 67–77.

Cerutti, L., Rodriguez, J.B., Tournie, E., 2010. GaSb-based laser, monolithically grown on silicon substrate, emitting at 1.55 μm at room temperature. IEEE Photonics Technol. Lett. 22 (8), 553–555.

Chen, G., Chen, H., Haurylau, M., Nelson, N.A., Albonesi, D.H., Fauchet, P.M., Friedman, E.G., 2007. Predictions of CMOS compatible on-chip optical interconnect. Integr. VLSI J. 40 (4), 434–446.

Doany, F., 2012. In: Power-efficient, high-bandwidth optical interconnects for high performance computing. IEEE Hot Interconnects Symposium, Santa Clara, CA, August 21–23, 2012.

Fang, A.W., Park, H., Cohen, O., Jones, R., Paniccia, M.J., Bowers, J.E., 2006. Electrically pumped hybrid AlGaInAs-silicon evanescent laser. Opt. Express 14 (20), 9203–9210.

Fang, A.W., Jones, R., Park, H., Cohen, O., Raday, O., Paniccia, M.J., Bowers, J.E., 2007. Integrated AlGaInAs-silicon evanescent race track laser and photodetector. Opt. Express 15 (5), 2315–2322.

Fang, A.W., Lively, E., Kuo, Y.-H., Liang, D., Bowers, J.E., 2008. A distributed feedback silicon evanescent laser. Opt. Express 16 (7), 4413–4419.

Fujii, T., Takeda, K., Diamantopoulos, N.-P., Kanno, E., Hasebe, K., Nishi, H., Nakao, R., Kakitsuka, T., Matsuo, S., 2018. Heterogeneously integrated membrane lasers on Si substrate for low-operating energy optical links. IEEE J. Sel. Top. Quantum Electron. 24 (1), 1500408.

Fujita, M., Ushigome, R., Baba, T., 2000. Continuous wave lasing in GalnAsP microdisk injection laser with threshold current of 40 μA. Electron. Lett. 36 (9), 790–791.

Futami, M., Shindo, T., Koguchi, T., Shinno, K., Amemiya, T., Nishiyama, N., Arai, S., 2012. GaInAsP/InP lateral current injection laser with uniformly distributed quantum-well structure. IEEE Photon. Technol. Lett. 24 (11), 888–890.

Goodman, J., Leonberger, F.I., Kung, S., Athale, R.A., 1984. Optical interconnections for VLSI systems. Proc. IEEE 72 (7), 850–866.

Gu, Z., Inoue, D., Amemiya, T., Nishiyama, N., Arai, S., 2018. 20 Gbps operation of membrane-based GaInAs/InP waveguide-type p–i–n photodiode bonded on Si substrate. Appl. Phys. Express 11 (2), 22102.

Heck, M.J.R., Chen, H.-W., Fang, A.W., Koch, B.R., Liang, D., Park, H., Sysak, M.N., Bowers, J.E., 2011. Hybrid silicon photonics for optical interconnects. IEEE J. Sel. Top. Quantum Electron. 17 (2), 333–346.

Heck, M.J.R., Bauters, J.F., Davenport, M.L., Doylend, J.K., Jain, S., Kurczveil, G., Srinivasan, S., Tang, Y., Bowers, J.E., 2013. Hybrid silicon photonic integrated circuit technology. IEEE J. Sel. Top. Quantum Electron. 19 (4), 6100117.

Hiratani, T., Shindo, T., Doi, K., Atsuji, Y., Inoue, D., Amemiya, T., Nishiyama, N., Arai, S., 2015a. Energy cost analysis of membrane distributed-reflector lasers for on-chip optical interconnects. IEEE J. Sel. Top. Quantum Electron. 21 (6), 299–308.

Hiratani, T., Inoue, D., Tomiyasu, T., Atsuji, Y., Fukuda, K., Amemiya, T., Nishiyama, N., Arai, S., 2015b. Room-temperature continuous-wave operation of membrane distributed-reflector laser. Appl. Phys. Express 8 (11), 112701.

Ho, R., Mai, K., Horowitz, M., 2001. The future of wires. Proc. IEEE 89 (4), 490–504.

Iga, K., 2000. Surface-emitting laser—its birth and generation of new optoelectronics field. IEEE J. Sel. Top. Quantum Electron. 6 (6), 1201–1215.

Inoue, D., Hiratani, T., Atsuji, Y., Tomiyasu, T., Amemiya, T., Nishiyama, N., Arai, S., 2015. Monolithic integration of membrane-based butt-jointed built-in DFB lasers and p-i-n photodiodes bonded on Si substrate. IEEE J. Sel. Top. Quantum Electron. 21 (6), 1502907.

Jeong, K.Y., No, Y.S., Hwang, Y., Kim, K.S., Seo, M.K., Park, H.G., Lee, Y.H., 2013. Electrically driven nanobeam laser. Nat. Commun. 4, 2822.

Kang, Y., Liu, H.D., Morse, M., Paniccia, M.J., Zadka, M., Litski, S., Sarid, G., Pauchard, A., Kuo, Y.H., Chen, H.W., Zaoui, W.S., Bowers, J.E., Beling, A., McIntosh, D.C., Zheng, X., Campbell, J.C., 2009. Monolithic germanium/silicon avalanche photodiodes with 340 GHz gain-bandwidth product. Nat. Photonics 3 (1), 59–63.

Liang, D., Fiorentino, M., Okumura, T., Chang, H.-H., Spencer, D.T., Kuo, Y.-H., Fang, A.W., Dai, D., Beausoleil, R.G., Bowers, J.E., 2009. Electrically-pumped compact hybrid silicon microring lasers for optical interconnects. Opt. Express 17 (22), 20355–20364.

Liao, L., Liu, A., Rubin, D., Basak, J., Chetrit, Y., Nguyen, H., Cohen, R., Izhaky, N., Paniccia, M., 2007. 40 Gbit/s silicon optical modulator for high-speed applications. Electron. Lett. 43 (22), 1196–1197.

Liebich, S., Zimprich, M., Ludewig, P., Beyer, A., Volz, K., Stolz, W., Kunert, B., Hossain, N., Jin, S.R., Sweeney, S.J., 2010. In: MOVPE growth and characterization of Ga(NAsP) laser structures monolithically integrated on Si (001) substrates. Proceedings of the 22nd IEEE International Semiconductor Laser Conference WA5, pp. 143–144.

Liu, J., Sun, X., Kimerling, L.C., Michel, J., 2009. Direct-gap optical gain of Ge on Si at room temperature. Opt. Lett. 34 (11), 1738–1740.

Liu, J., Sun, X., Aguilera, R.C., Kimerling, L.C., Michel, J., 2010. Ge-on-Si laser operating at room temperature. Opt. Lett. 35 (5), 679–681.

Maekawa, T., Ito, H., Masu, K., 2008. In: An 8 Gbps 2.5 mW on-chip pulsed-current-mode transmission line interconnect with a stacked-switch Tx. Proceedings of the 34th European Solid-State Circuits Conference, pp. 474–477.

Manipatruni, S., Chen, L., Lipson, M., 2009. In: 50 Gbit/s wavelength division multiplexing using silicon microring modulators. IEEE International Conference on Group IV Photonics GFP, FC3.

Maruyama, T., Okumura, T., Sakamoto, S., Miura, K., Nishimoto, Y., Arai, S., 2006. GaInAsP/InP membrane BH-DFB lasers directly bonded on SOI substrate. Opt. Express 14 (18), 8184–8188.

Matsuo, S., Shinya, A., Kakitsuka, T., Nozaki, K., Segawa, T., Sato, T., Kawaguchi, Y., Notomi, M., 2010. High-speed ultracompact buried heterostructure photonic-crystal laser with 13 fJ of energy consumed per bit transmitted. Nat. Photonics 4 (9), 648–654.

Mensink, E., Schinkel, D., Klumperink, E., Van Tuijl, E., Nauta, B., 2007. In: A 0.28 pJ/b 2 Gb/s/ch transceiver in 90 nm CMOS for 10 mm on-chip interconnects. IEEE International Solid-State Circuits Conference, pp. 414–416.

Miller, D.A.B., 2000. Rationale and challenges for optical interconnects to electronic chips. Proc. IEEE 88 (6), 728–749.

Miller, D.A.B., 2009. Device requirements for optical interconnects to silicon chips. Proc. IEEE 97 (7), 1166–1185.

Naitoh, H., Sakamoto, S., Ohtake, M., Okumura, T., Maruyama, T., Nishiyama, N., Arai, S., 2007. GaInAsP/InP membrane buried heterostructure distributed feedback laser with air-bridge structure. Jpn. J. Appl. Phys. Part 2 Lett. 46 (47), L1158–L1160.

Oe, K., Noguchi, Y., Caneau, C., 1994. GaInAsP lateral current injection lasers on semi-insulating substrates. IEEE Photonics Technol. Lett. 6 (4), 479–481.

Ohashi, K., Nishi, K., Shimizu, T., Nakada, M., Fujikata, J., Ushida, J., Toru, S., Nose, K., Mizuno, M., Yukawa, H., Kinoshita, M., Suzuki, N., Gomyo, A., Ishi, T., Okamoto, D., Furue, K., Ueno, T., … Akedo, J., 2009. On-chip optical interconnect. Proc. IEEE 97 (7), 1186–1196.

Okamoto, T., Nunoya, N., Onodera, Y., Tamura, S., Arai, S., 2002. Low-threshold singlemode operation of membrane BH-DFB lasers. Electron. Lett. 38 (23), 1444–1446.

Okamoto, T., Nunoya, N., Onodera, Y., Yamazaki, T., Tamura, S., Arai, S., 2003. Optically pumped membrane BH-DFB lasers for low-threshold and single-mode operation. IEEE J. Sel. Top. Quantum Electron. 9 (5), 1361–1366.

Okamoto, T., Yamazaki, T., Sakamoto, S., Tamura, S., Arai, S., 2004. Low-threshold membrane BH-DFB laser arrays with precisely controlled wavelength over a wide range. IEEE Photonics Technol. Lett. 16 (5), 1242–1244.

Okumura, T., Maruyama, T., Kanemaru, M., Sakamoto, S., Arai, S., 2007. Single-mode operation of GaInAsP/InP-membrane distributed feedback lasers bonded on silicon-on-insulator substrate with rib-waveguide structure. Jpn. J. Appl. Phys. 46 (48), l1206–L1208.

Okumura, T., Maruyama, T., Yonezawa, H., Nishiyama, N., Arai, S., 2009. Injection-type GaInAsP-InP-Si distributed-feedback laser directly bonded on silicon-on-insulator substrate. IEEE Photonics Technol. Lett. 21 (5), 283–285.

Okumura, T., Ito, H., Kondo, D., Nishiyama, N., Arai, S., 2010. Continuous wave operation of thin film lateral current injection lasers grown on semi-insulating InP substrate. Jpn. J. Appl. Phys. 49 (4), 40205.

Okumura, T., Koguchi, T., Ito, H., Nishiyama, N., Arai, S., 2011. Injection-type GaInAsP/InP membrane buried heterostructure distributed feedback laser with wire like active regions. Appl. Phys. Express 4 (4), 42101.

Park, H.-G., Kim, S.-H., Kwon, S.-H., Ju, Y.-G., Yang, J.-K., Baek, J.-H., Kim, S.-B., Lee, Y.-H., 2004. Electrically driven single-cell photonic crystal laser. Science 305 (5689), 1444–1447.

Rattner, J., 2006. The Mega-Center. Intel Developer Forum.

Razeghi, M., Defour, M., Blondeau, R., Omnes, F., Maurel, P., Acher, O., Brillouet, F., C-Fan, J.C., Salerno, J., 1988. First cw operation of a $Ga_{0.25}In_{0.75}As_{0.5}P_{0.5}$-InP laser on a silicon substrate. Appl. Phys. Lett. 53 (24), 2389–2390.

Roelkens, G., Van Thourhout, D., Baets, R., Nötzel, R., Smit, M., 2006. Laser emission and photodetection in an InP/InGaAsP layer integrated on and coupled to a silicon-on-insulator waveguide circuit. Opt. Express 14 (18), 8154–8159.

Rong, H., Jones, R., Liu, A., Cohen, O., Hak, D., Fang, A., Paniccia, M., 2005. A continuous-wave Raman silicon laser. Nature 433 (7027), 725–728.

Rong, H., Xu, S., Kuo, Y.H., Sih, V., Cohen, O., Raday, O., Paniccia, M., 2007. Low-threshold continuous-wave Raman silicon laser. Nat. Photonics 1 (4), 232–237.

Rouvìre, M., Vivien, L., Le Roux, X., Mangeney, J., Crozat, P., Hoarau, C., Cassan, E., Pascal, D., Laval, S., F′d′li, J.M., Damlencourt, J.F., Hartmann, J.M., Kolev, S., 2005. Ultrahigh speed germanium-on-silicon-on-insulator photodetectors for 1.31 and 1.55 μm operation. Appl. Phys. Lett. 87 (23), 231109.

Sakamoto, S., Okamoto, T., Yamazaki, T., Tamura, S., Arai, S., 2005. Multiple-wavelength membrane BH-DFB laser arrays. IEEE J. Sel. Top. Quantum Electron. 11 (5), 1174–1179.

Sakamoto, S., Kawashima, H., Naitoh, H., Tamura, S., Maruyama, T., Arai, S., 2007a. Reduced temperature dependence of lasing wavelength in membrane buried heterostructure DFB lasers with polymer cladding layers. IEEE Photonics Technol. Lett. 19 (5), 291–293.

Sakamoto, S., Naitoh, H., Ohtake, M., Nishimoto, Y., Tamura, S., Maruyama, T., Nishiyama, N., Arai, S., 2007b. Strongly index-coupled membrane BH-DFB lasers with surface corrugation grating. IEEE J. Sel. Top. Quantum Electron. 13 (5), 1135–1141.

Sakamoto, S., Naitoh, H., Ohtake, M., Nishimoto, Y., Maruyama, T., Nishiyama, N., Arai, S., 2007c. 85°C continuous-wave operation of GaInAsP/InP-membrane buried heterostructure distributed feedback lasers with polymer cladding layer. Jpn. J. Appl. Phys. 46 (47), L1155–L1157.

Shindo, T., Okumura, T., Ito, H., Koguchi, T., Takahashi, D., Atsumi, Y., Kang, J., Osabe, R., Amemiya, T., Nishiyama, N., Arai, S., 2011a. GaInAsP/InP lateral-current-injection distributed feedback laser with a-Si surface grating. Opt. Express 19 (3), 1884–1891.

Shindo, T., Okumura, T., Ito, H., Koguchi, T., Takahashi, D., Atsumi, Y., Kang, J., Osabe, R., Amemiya, T., Nishiyama, N., Arai, S., 2011b. Lateral-current-injection distributed feedback laser with surface grating structure. IEEE J. Sel. Top. Quantum Electron. 17 (5), 1175–1182.

Shindo, T., Futami, M., Doi, K., Amemiya, T., Nishiyama, N., Arai, S., 2013a. Design of lateral-current-injection-type membrane distributed-feedback lasers for on-chip optical interconnections. IEEE J. Sel. Top. Quantum Electron. 19 (4), 1502009.

Shindo, T., Futami, M., Okumura, T., Osabe, R., Koguchi, T., Amemiya, T., Nishiyama, N., Arai, S., 2013b. Lateral-current-injection type membrane DFB laser with surface grating. IEEE Photonics Technol. Lett. 25 (13), 1282–1285.

Spuesens, T., Liu, L., De Vries, T., Romeo, P.R., Regreny, P., Van Thourhout, D., 2009. In: Improved design of an InP-based microdisk laser heterogeneously integrated with SOI. IEEE International Conference on Group IV Photonics GFP FA3.

Sugo, M., Mori, H., Tachikawa, M., Itoh, Y., Yamamoto, M., 1990. Room-temperature operation of an InGaAsP double-heterostructure laser emitting at 1.55 μm on a Si substrate. Appl. Phys. Lett. 57 (6), 593–595.

Takeda, K., Sato, T., Shinya, A., Nozaki, K., Kobayashi, W., Taniyama, H., Notomi, M., Hasebe, K., Kakitsuka, T., Matsuo, S., 2013. Few-fJ/bit data transmissions using directly modulated lambda-scale embedded active region photonic-crystal lasers. Nat. Photonics 7 (7), 569–575.

Tomiyasu, T., Hiratani, T., Inoue, D., Nakamura, N., Fukuda, K., Uryu, T., Amemiya, T., Nishiyama, N., Arai, S., 2017. High-differential quantum efficiency operation of GaInAsP/InP membrane distributed-reflector laser on Si. Appl. Phys. Express 10 (6), 62702.

Tomiyasu, T., Inoue, D., Hiratani, T., Fukuda, K., Nakamura, N., Uryu, T., Amemiya, T., Nishiyama, N., Arai, S., 2018. 20-Gbit/s direct modulation of GaInAsP/InP membrane distributed-reflector laser with energy cost of less than 100 fJ/bit. Appl. Phys. Express 11 (1), 12704.

TSUBAME2.5, 2013. Hardware Software Specifications. Tokyo Institute of Technology. http://www.gsic.titech.ac.jp/en/node/420.

Van Campenhout, J., Rojo Romeo, P., Regreny, P., Seassal, C., Van Thourhout, D., Verstuyft, S., Di Cioccio, L., Fedeli, J.-M., Lagahe, C., Baets, R., 2007. Electrically pumped InP-based microdisk lasers integrated with a nanophotonic silicon-on-insulator waveguide circuit. Opt. Express 15 (11), 6744–6749.

Van Campenhout, J., Binetti, P.R.A., Romeo, P.R., Regreny, P., Seassal, C., Leijtens, X.J.M., De Vries, T., Oei, Y.S., Van Veldhoven, R.P.J., Nötzel, R., Di Cioccio, L., Fedeli, J.M., Smit, M.K., Van Thourhout, D., Baets, R., 2009. Low-footprint optical interconnect on an SOI chip through heterogeneous integration of InP-based microdisk lasers and microdetectors. IEEE Photonics Technol. Lett. 21 (8), 522–524.

Vivien, L., Osmond, J., Fédéli, J.-M., Marris-Morini, D., Crozat, P., Damlencourt, J.-F., Cassan, E., Lecunff, Y., Laval, S., 2009. 42 GHz pin germanium photodetector integrated in a silicon-on-insulator waveguide. Opt. Express 17 (8), 6252–6257.

Wada, H., Kamijoh, T., 1996. Room-temperature CW operation of InGaAsP lasers on Si fabricated by wafer bonding. IEEE Photonics Technol. Lett. 8 (2), 173–175.

Yang, G.M., MacDougal, M.H., Dapkus, P.D., 1995. Ultralow threshold current vertical-cavity surface-emitting lasers obtained with selective oxidation. Electron. Lett. 31 (11), 886–888.

Yin, T., Cohen, R., Morse, M.M., Sarid, G., Chetrit, Y., Rubin, D., Paniccia, M.J., 2007. 31 GHz Ge n-i-p waveguide photodetectors on silicon-on-insulator substrate. Opt. Express 15 (21), 13965–13971.

Zheng, X., Lexau, J., Luo, Y., Thacker, H., Pinguet, T., Mekis, A., Li, G., Shi, J., Amberg, P., Pinckney, N., Raj, K., Ho, R., Cunningham, J.E., Krishnamoorthy, A.V., 2010. Ultra-low-energy all-CMOS modulator integrated with driver. Opt. Express 18 (3), 3059–3070.

CHAPTER FIVE

Photonic Crystal Lasers and Nanolasers on Silicon

Dimitris Fitsios*, Fabrice Raineri*,†,1
*Centre de Nanosciences et de Nanotechnologies, CNRS, Université Paris Saclay, Palaiseau, France
†Université Paris Denis Diderot, Sorbone Paris Cité, Paris, France
[1]Corresponding author: e-mail address: fabrice.raineri@c2n.upsaclay.fr

Contents

1. INTRODUCTION

1.1 What Is a Photonic Crystal

Photonic crystal (PhC) structures have long been in the spotlight of laser research, due to their enticing characteristics that can enable extreme tailoring of the electromagnetic (EM) field, applicable for the thorough control of the spontaneous emission of a laser. The term "photonic crystal" was initially introduced during the 1980s by Yablonovitch (1987) and John (1987). Yablonovitch's work revolved around the inhibition of spontaneous emission in a wavelength-scale periodic refractive index modulation in space

Semiconductors and Semimetals, Volume 99
ISSN 0080-8784
https://doi.org/10.1016/bs.semsem.2018.08.003

(Yablonovitch, 1987), while John worked on its acceleration (John, 1987). The PhC structures were inspired by atomic lattices, having spatially varying dielectric constant, as atomic lattices have the atomic potential. Thus, a significant feature of PhC structures is the photonic band gap (PBG), where propagation of light is blocked for specific wavelength bands.

It comes without doubt that one of the primary targets when developing lasers is the absolute minimization of the lasing threshold. To achieve that, of course, resonating structures of high quality are needed. PhCs are among the best in this category, as they exhibit record Q/V ratios, where Q is the quality factor and V the modal volume of the resonant mode.

This chapter is dedicated to PhC nanolasers, and particularly those integrated on silicon (Si) boards. Until recently, PhC nanolasers were at an infant level, having still several issues to overcome before making the next step to become a force to be reckoned in the commercial world. However, lately significant progress has been made, indicating perhaps that it is the right time for PhC nanolasers to make this next step. In this chapter, we revisit the roadmap that led to this evolution and analyze the main milestones behind it.

This chapter is divided into four parts:

The first one deals with the main guidelines that need to be taken into account when designing a nanocavity and the issues that need to be overcome to develop high-performance and efficient PhC nanolasers. In the second part, we will analyze the design and performance of optically pumped PhC nanolasers on Si or silicon-on-insulator (SOI) wafers and, in specific, PhC nanolasers on blank Si—no waveguides—and SOI circuitry, which can serve as the basis for the solving of several issues. The third part will review the latest research efforts on electrically injected PhC nanolasers on Si/SOI circuitry, which have solved all the major issues and promise to bring PhC nanolasers on the next crest of the wave of commercial CMOS integrated nanophotonics.

Finally, the fourth part will summarize and provide examples for further manipulation of PhC nanocavities in alternative structures.

1.2 How to Construct a PhC Nanocavity

As with other photonic structures, a good way to properly determine the optical properties of PhCs is the definition of the photonic band structure. In the diagram representing the photonic band structure, the optical mode frequencies are plotted as a function of the wavevector (Joannopoulos et al., 2008). By adjusting the architectural parameters of the PhC, such as the lattice constant or the radius of the holes, the photonic band structure can be tuned to the absolute targeted user specifications. The photonic band structure of a two-dimensional PhC (hexagonal lattice of air holes drilled into the material) for the TM/TE polarization states can be seen in Fig. 1.

The first thing that could catch the eye is the presence of a "gap" inside the photonic band structure. In this gap, the PBG, for a range of frequencies light is forbidden to propagate in the considered direction. These structures are the photonic analogue of atomic lattices, where the spatially varying dielectric constant ($\varepsilon = n^2$) plays the role of the atomic potential. The PBG is created due to the constructive and destructive interferences that occur inside the wavelength-scale periodic structure. As already mentioned,

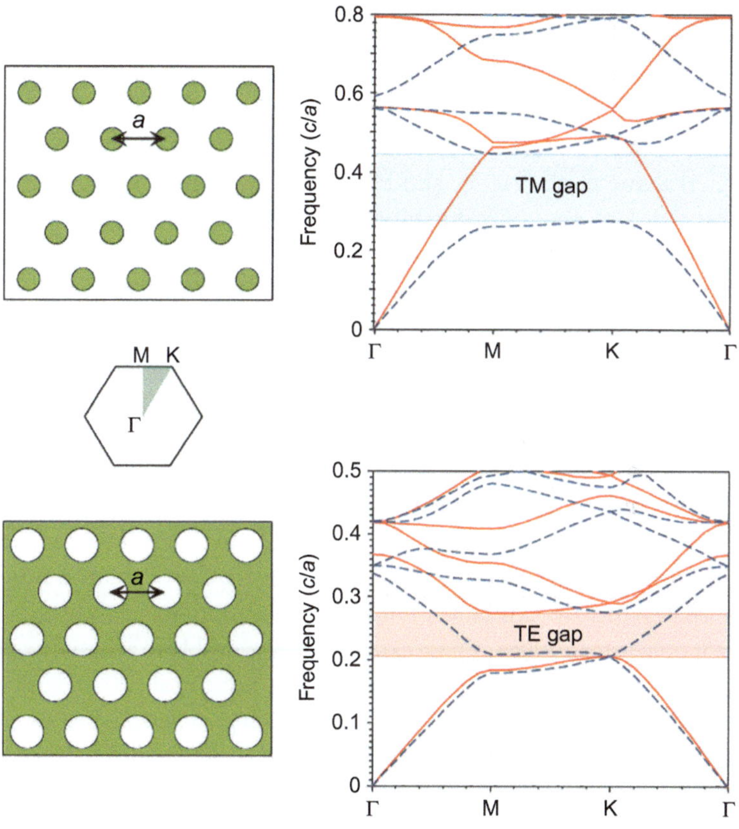

Fig. 1 Band diagrams and photonic band gaps for hexagonal lattices of high dielectric rods in air (*top*) and air holes in dielectric (*bottom*), where *a* is the center–center periodicity. The frequencies are plotted around the boundary of the irreducible Brillouin zone (*shaded triangle, left center*) (Johnson, 2003).

by varying the optogeometrical parameters of the PhC, we can alter features such as the dispersion and engineer the PBG to our liking. As such, we are able to obtain PBGs that can be utilized to confine light within line-defect waveguides or point-defect cavities, as we will discuss further in this chapter. Moreover, we can make use of the "band-edge" frequencies, where the bands flatten when the wavevector corresponds to the high symmetry points of the first Brillouin zone, which can be utilized to obtain extremely low group velocity (v_g) values (Sakoda, 2005).

In this chapter, we will focus on PhCs that use III–V semiconductors as the main material, providing the required gain medium for the fabrication of PhC nanolasers. The two main categories of III–V compounds that have been used so far are quantum dots (QDs) and quantum wells (QWs). Taking advantage of modern technologies of lithography and etching, they can potentially be turned into efficient lasing structures. Lasers based on QDs have a respectable part in the literature, however fall outside the scope of this chapter. If the reader wants to pursue further education on the matter, more information can be found in Nomura et al. (2006, 2007), Kounoike et al. (2005), Akahane et al. (2003), and Hostein et al. (2010). QWs are possibly the most standard way of inducing gain potential in PhC structures. They have been employed in semiconductor membranes with etched PhC architecture to fabricate different types of lasers, such as coupled-cavity PhC lasers (Huang et al., 2001), 1D nanocavity lasers (Halioua et al., 2009, 2011), and buried heterostructure nanolasers (Matsuo et al., 2012; Takeda et al., 2017).

The two major types of achieved PhC lasers have been evident in the literature so far.

1.2.1 Nanocavity (1D) PhC Lasers

In order to engineer a PBG in the nanocavity, a defect is introduced in the periodic lattice, creating new states within the PBG at certain frequencies, which correspond to spatially confined optical modes within the defect (Joannopoulos et al., 2008). These cavities have been able to achieve extremely high Q/V ratio values with Q factors above 10^6 (Asano et al., 2006; Kuramochi et al., 2006). Nanobeam cavities fall into this category, consisting of a single row of holes drilled into a semiconductor ridge waveguide (Foresi et al., 1997; Halioua et al., 2010), on top of silicon substrates and potentially bonding media such as BCB, which can be seen in Fig. 2. The defect in the center of the cavity can be easily seen, as well as the tapering and mirror regions.

Although nanobeam cavities provide a vast reduction in footprint, with values as low as $10\,\mu m^2$, their performance can reach values similar to that of

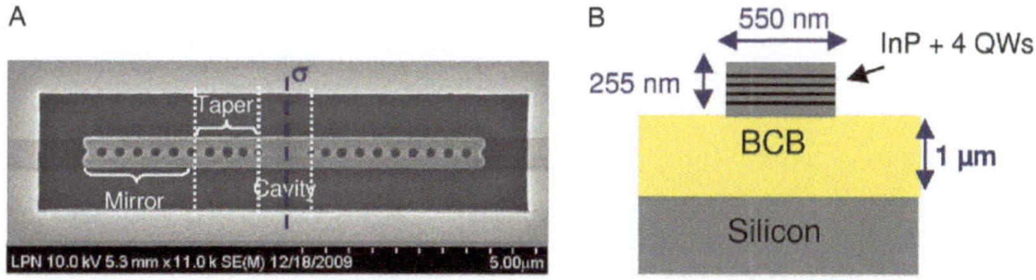

Fig. 2 Wire cavity laser. (A) SEM picture of the top view of the wire cavity laser. (B) Schematics of the transverse profile of the wire cavity (Halioua et al., 2010).

2D PhC structures in terms of Q/V ratios. Moreover, the fact that nanobeam cavities can be easily integrated having SiO_2 as the surrounding material instead of air—providing efficient heat sinking and mechanical stability—and still provide competitive performance has led to a series of laser demonstrations with promising results (Halioua et al., 2010; Vuckovic et al., 2002; Zhang et al., 2010). Such a device will be analyzed later in this chapter.

1.2.2 Lasers Based on 2D PhC Structures

2D PhC structures are probably the most common PhC architectures. An example can be seen in Fig. 3A. Again, we can observe the periodic arrangement of air holes etched into the semiconductor, which can be surrounded by a low-index material such as air or silica. The PBG, as in the case of 1D PhC structures, can be engineered by varying the lattice constant across certain regions to create a defect, as can be seen in Fig. 3A and B, or by varying the radius of the air holes, respectively. If we combine these effects with the adjustment of propagation of light induced by the stack of layers in the direction normal to the plane of periodicity forming the waveguide, we can achieve confinement of light with the combined effect of the PBG and total internal reflection at the interface with the low-index material.

1.3 Basic Characteristics of PhC Nanolasers

PhC nanolasers, besides their compact footprint, have stood out in the laser family due to their two important characteristics:

- Low threshold or even threshold-less operation
- High-speed operation

Plenty of effort has been devoted to describe the static and dynamic characteristics of PhC semiconductors and analyze their particularities in terms of analytics equations. One could say that this framework is close to completion and standard rate equations can largely describe the main characteristics of PhC nanolasers. As it falls outside the scope of this chapter, we will

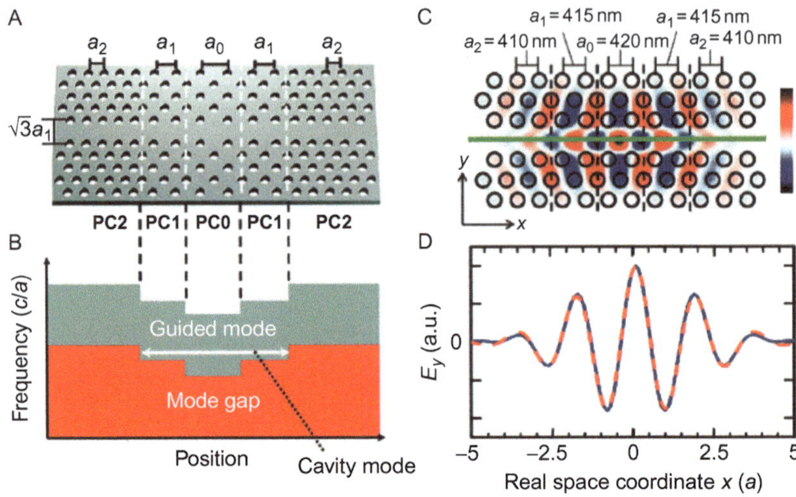

Fig. 3 (A) Schematic image of a 2D PhC multiheterostructure nanocavity. (B) Respective band diagram along the waveguide direction. Photons of a specific energy (*white arrow*) can exist only in the waveguide of PC0 and PC1. (C) Calculated E_y distribution in the multiheterostructure nanocavity and (D) E_y profile along the waveguide direction (Tanaka et al., 2008).

not go into detail on the equation models, which anyway can be easily found in the literature (Coldren et al., 2012; Englund et al., 2007a,b; Raineri et al., 2014; Xu et al., 2000).

Nevertheless, it is important to note that one of the most impressive features of PhC nanolasers is the inherent possibility they provide to control spontaneous emission, providing either inhibition or acceleration of it (Badolato et al., 2005; Englund et al., 2005; Fujita et al., 2005; Laurent et al., 2005). This property greatly differentiates them from lasers built from bulk materials. In other words, the radiative lifetime of a PhC nanolaser can be modified by taking advantage of the Purcell effect (Purcell et al., 1946). As such, it is governed by the Purcell factor, F_p:

$$\frac{F_p}{\tau_{r\_bulk}} = \frac{1}{\tau_r}$$

connecting thus τ_r, the radiative lifetimes in PhC and τ_{r_bulk}, in bulk structures. The Purcell factor can found as

$$F_p = \frac{3}{4\pi^2}\left(\frac{\lambda}{n}\right)^3 \frac{Q_p}{V}$$

where λ is the wavelength, n is the refractive index, V is the modal volume, and Q_p is the quality factor that is altered by the cavity photon lifetime τ_p and dephasing time T_2 by (Xu et al., 2000):

$$\frac{1}{Q_p}=\frac{1}{\omega}\left(\frac{1}{\tau_p}+\frac{1}{T_2}\right)$$

One could easily notice the dependence on the Purcell factor on the modal volume, which is an inherent advantage of PhCs with their modal volumes near $(\lambda/n)^3$, giving Purcell factor values close to or even larger than 1. Of course, Q_p should be always large enough, depending on τ_p and T_2. It has been shown that T_2 can reach values down to tens or hundreds of ps when at cryogenic temperatures (4°K) (Laurent et al., 2005), only to be limited by the photon lifetime. Purcell factor values of 20 have been demonstrated experimentally for such cases (Laurent et al., 2005). At room temperatures, T_2 can be in the order of tens of fs and much shorter than the photon lifetime, giving a value for the Purcell factor around 1.

Meanwhile, the control of the spontaneous emission strongly affects the spontaneous emission factor β, which is the ratio of the laser mode emitted photons to the total emitted photons. In PhC nanolasers, β factor can be really maximized compared to conventional lasers. This happens as in PhC nanolasers, the spectral width of the emitter is much narrower than that of the cavity and the larger part light is emitted in the directions that couple to the optical mode. The two situations are visualized in Fig. 4.

If we try to describe β, we get

$$\beta=\frac{\Gamma_{\text{mode}}}{\Gamma_{\text{total}}}=\frac{F_p}{F_p+\gamma}$$

where Γ_{mode} is the spontaneous emission rate in the cavity mode and Γ_{total}, the total spontaneous emission rate. γ is the ratio between the rate of spontaneous emission in all other modes over the rate of spontaneous emission if the emitters were in the bulk material. We saw before how F_p can reach increased values in PhCs; however, γ has an even more astonishing effect. When we are dealing with PhCs, we can have inhibition of the spontaneous emission that is not coupled into the cavity mode by taking advantage of a property that has been described before, the PBG. Consequently, γ factor can reach even values tending to 0, in the case that we operate at cryogenic temperatures, where we can theoretically obtain threshold-less operation. The case of cryogenic operation and a high β value of 0.85 can

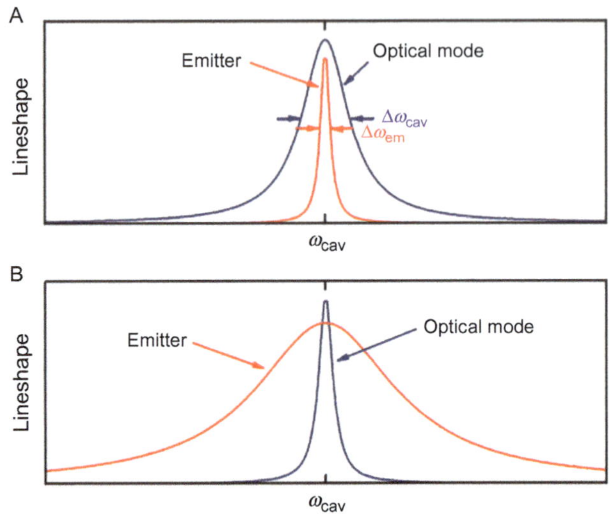

Fig. 4 (A) Cavity mode resonance that is broader than the emitter's transition width. (B) Cavity mode resonance that is narrower than the emitter's transition width (Romeira and Fiore, 2018).

be seen in Fig. 5A. Even at room temperature, γ reaches a value of around 0.1, rendering β orders of magnitude greater than that of conventional lasers. β in PhC nanolasers can theoretically even approach values near 1, whereas the respective values for edge-emitting bulk lasers are in the range of 10^{-5}–10^{-4} (Coldren et al., 2012) and in the range of 10^{-5}–10^{-2} for VCSELs (Koyama et al., 1991; Kuksenkov et al., 1997). This fact greatly contributes to the extremely reduced threshold of PhC nanolasers. The effect of the improved β value can be seen in Fig. 5B, having reduced threshold for higher β values. It should be noted that a very recent work has also pointed out that ultrafast modulation-speed properties are predicted in nanolasers as a direct consequence of the enhancement of the stimulated emission rate via reduction of the mode volume, attributing to them another significant advantage (Romeira and Fiore, 2018).

1.4 Optimization of 2D PhC Structures

As with every resonating structure used for building efficient lasers, PhC structures need to have maximized Q factors, while keeping modal volumes minimized. To design such cavities, it is imperative to handle precisely the losses of the cavity, both in-plane and out-of-plane. In-plane losses arise

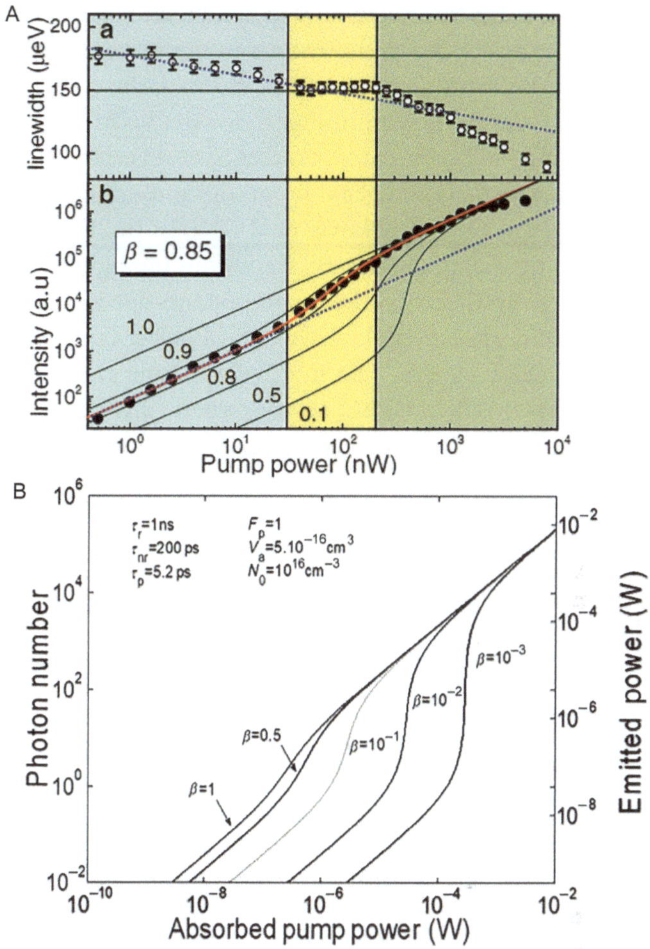

Fig. 5 (A) Light–light curve measured at 4 K on a 2D PhC cavity. (B) Light–light characteristics of PhC lasers with different values of β (Strauf et al., 2006).

from the number and type of holes present on the sides of the cavity and can be directly manipulated by adjusting the number and type of holes. On the other hand, out-of-plane losses may arise from light diffracting at the interface between the defect and the mirror regions. Moreover, in the third direction, light is confined by exploiting total internal reflection. However, for this to be valid, the "light cone" relation has to be satisfied (John, 1987). As such, the modulus of the in-plane wavevector $k_{||}$ must obey the following: $k_{||} > 2\pi n_s/\lambda$, where λ is the wavelength, and n_s the refractive index of the surrounding medium.

Fourier transform can be a useful tool to handle the electromagnetic field amplitude distribution for the cavity modes in the wavevector space. This can be achieved by performing a Fourier transform of the spatial distribution just above the slab, as can be seen in Fig. 6, for the spatial and wavevector distributions of a 2D PhC cavity in Si. Studies have revealed a strong dependence of the Q factor of 2D PhC cavities on the amount of the field distribution inside the light cone (Vuckovic et al., 2002). In 2D PhC cavities, a significant part of the field components lie within the light cone. Thus, in order to maximize the Q factor, these components need to be minimized (Akahane et al., 2003; Asano et al., 2006), and Fourier transform can serve as the means for achieving this. The study of Akahane et al. (2003) can be seen in Fig. 6, for an L3 2D PhC cavity, where three holes have been removed from the lattice at the defect region. The light cone is highlighted

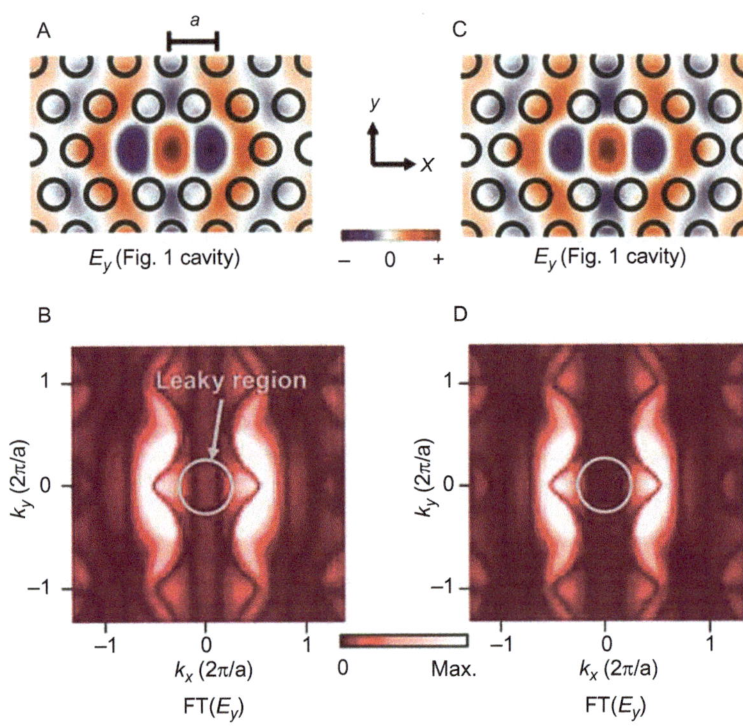

Fig. 6 (A) Spatial distribution of the E_y field amplitude for the initial L3 cavity. (B) Wavevector-space distribution of the E_y field amplitude for the initial L3 cavity. (C) Distribution of the E_y field amplitude for the modified L3 cavity. (D) Wavevector-space distribution of the E_y field amplitude for the modified L3 cavity (Akahane et al., 2003).

with the white circle and signifies the leaky region for the field distribution coupling to the continuum of radiative modes. As we can see, a big part of the field distribution lies within the light cone. However, by shifting the two inner holes of the cavity, the part of the field components lying within the light cone gets greatly reduced, greatly improving the overall Q factor.

More thorough guidelines on the design of 2D PhC cavities with high Q factor can be found in the literature. As it falls outside the exact scope of this chapter, below we provide a brief summary of some of the most basic points in them (Akahane et al., 2003; Asano et al., 2006; Quan and Loncar, 2011; Velha et al., 2006):

- Abrupt changes in the field amplitude spatial envelope generate large field components in the leaky region, thus hampering the Q factor. The study in Akahane et al. (2003) points out that the implementation of the so-called gentle confinement can reduce the amplitude of these components. Specifically, the field spatial envelope has a Gaussian profile, thus forcing a similar envelope for the wavevector, decaying rapidly, and vanishing quasi-completely inside the light cone. Several such PhC structures have been proposed, both 1D (Halioua et al., 2011) and 2D (Akahane et al., 2003; Song et al., 2005), while a structure worth pointing out is the multiheterostructure. In the multiheterostructure cavity, a whole row of holes is missing, thus forming a PhC waveguide, and the cavity is generated by varying the lattice constant along the propagation direction, gradually increasing the mirror strength. The mirror strength is actually the imaginary part of the propagation constant, thus achieving a field envelope with a Gaussian profile. The number of gradually increasing lattice constant mirror regions can of course be adjusted.
- A tapering region in the PhC lattice can create a "bridge" between the cavity and mirror evanescent modes, thus reducing the large losses that can be inflicted due to modal or impedance mismatch. As it has been shown (Lalanne and Hugonin, 2003; Sauvan et al., 2005), this can be greatly beneficial as it can greatly reduce scattering at the interfaces.
- For years, the use of air-bridged membranes has been very popular among PhC cavities, in order to reduce the size of the leaky region in the wavevector space. The aim is to take advantage of surrounding the PhC cavity with a medium that has a refractive index as small as possible and air serves perfectly as such a medium. However, even though the presence of an air bridge greatly enhances the Q factor, it renders the structure mechanically too unstable and prone to environmental conditions. Moreover, it falls outside the typical CMOS processes, raising

dangerously the degree of complications in fabrication. Consequently, alternative endeavors have been made to have enhanced Q factors while avoiding air bridge mechanisms. Such are, as we have said, the multiheterostructures with a Gaussian field profile which can lead to Q factor values over 10^7. Moreover, other techniques include the use of double heterostructure cavities (Braive et al., 2009; Song et al., 2005), the use of QDs (Hostein et al., 2010), or the use of 1D PhC nanowire cavities (Foresi et al., 1997). In the latter, the use of a semiconductor ridge waveguide with a single row of holes can greatly reduce the overall footprint while keeping constant with Q/V values of 2D PhC structures. Probably the most handy feature of these structures is the ability to have high Q factors without the presence of an air bridge, with the surrounding material being, e.g., silica. More on this technique can be found later on in this chapter.

1.5 Remaining Issues to Be Solved Before Mass Production

It is without doubt that PhC nanolasers have made tremendous progress over the last decades. What seemed to be exotic in the past now seems pragmatic, with the demonstration of several PhC nanolasers with low threshold, small footprint, and improvements over thermal issues. Still, however, steps need to be made to take PhC nanolasers to the next level, which is mass production and utilization in real applications. They need to be taken out of the research laboratory and into Data Centers or other computing, networking, or other real-life environments. To achieve that, efforts need to be focused on specific targets:

- In order to make the transition from single elements to large-scale photonic circuits, as well as to have easy interfacing with electronic counterparts on the CMOS board, electrical injection needs to be a standard for PhC nanolasers, with sophisticated electrical contacting solutions that do not hamper the optical properties of devices.
- The output power of the nanolaser needs to be directly and easily utilized by the existing photonic circuits on the Si board. Thus, simple, efficient, and high-yield coupling schemes need to be employed to channel the nanolaser output signal to the Si board.
- PhCs have an inherent flaw of surface nonradiative carrier recombination due to the holes exposing a large surface. Actions such as specific selection of materials and surface passivation need to be made to tackle this issue.

- Together with nonradiative carrier recombination, heat sinking is crucial in order to achieve CW and room-temperature (RT) operation of the devices, which is of course the only way for them to be practically useful in modern application systems. Air-bridged membranes provide increased confinement of the EM field due to the large index contrast of the active material with air vertically and horizontally. However, due to poor thermal conductivity of air and mechanical instability, other choices of surrounding material need to be made.

2. OPTICALLY PUMPED PhC NANOLASERS ON Si/SOI

There have been numerous research demonstrations of PhC nanolasers on Si/SOI employing optical pumping schemes over the years. In this chapter, we will visit some of them that stand out and better highlight the evolution over the years and the points that were improved and those that need further improvement to reach the operational levels that are needed for application-level performance.

2.1 PhC Nanolasers on Si Wafers

2.1.1 1D PhC Single-Mode Wire Waveguide on Si Wafer

A typical example of PhC nanolasers and a representative of the wire cavity nanolaser family were reported in Halioua et al. (2009, 2010, 2011). The device can be seen in Fig. 7. It is a 1D PhC single-mode wire waveguide with three regions: the defect region at the center of the waveguide without any holes, a mirror region on each side of the waveguide, and a tapering region at each defect–mirror interface formed with drilled air holes on a single row. As we have discussed previously, this type of nanocavities can lead to extremely high Q/V ratios, without having to resort to air bridge architectures. This device takes advantage of this fact, having a low-index ($n = 1.54$ at $1.55\,\mu m$) BCB (benzocyclobutene) of $1\,\mu m$ thickness as the bonding layer between the cavity and an unpatterned Si wafer. The cavity consists of a 550-nm-wide and 255-nm-thick InP strip waveguide incorporating four InGaAs/InGaAsP QWs. The two sets of holes at the extremities form mirrors composed by a periodic section of eight holes, while each tapering region comprises three holes with gradually decreasing diameters. Light is confined laterally within the cavity by index guiding due to the semiconductor/air interface, and in the vertical direction by air above and BCB below.

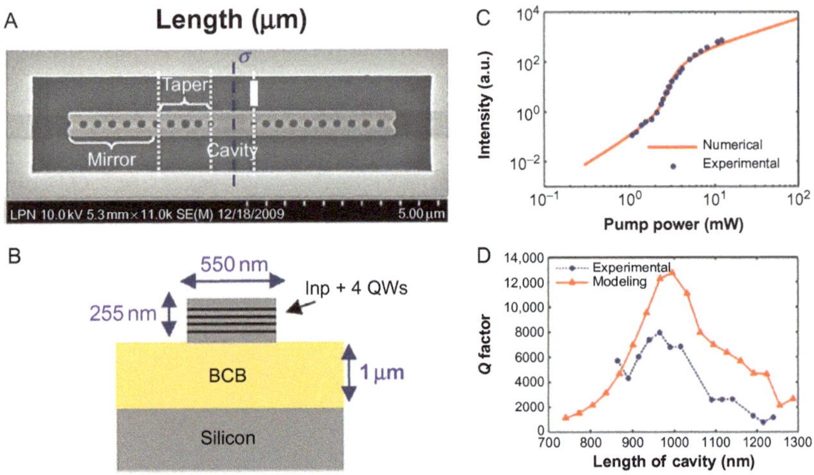

Fig. 7 (A) SEM image of 1D PhC single-mode wire nanolaser on Si wafer and (B) schematic of its vertical heterostructure. (C) Measured threshold and corresponding fitted rate equation vs pump power. (D) Experimental *Q* factors and modeled *Q* factors vs cavity length (Halioua et al., 2010).

A sensible question regarding this type of nanocavities would concern which factor defines the resonant wavelength of the cavity. The answer would be several factors: the length of the cavity, the effective index of the mode confined in the wire, and the phase change induced by the mirror reflection. Moreover, as we have discussed earlier in this chapter, a tapering region can reduce the large losses that can be inflicted due to modal or impedance mismatch. This is the case also for this device, as the tapering region adiabatically minimizes the modal mismatch between the guided mode and the evanescent mirror modes, decreasing the losses at the cavity–mirror interface, thereby increasing the reflectivity, and significantly enhancing the Q factor. Furthermore, the radius of the holes and the lattice constant are adjusted to reveal the largest bandgap, which in turn gives the best confinement, further enhancing the Q factor.

The device was operated at RT, by using an optical pump powered by a laser source at 808 nm, and modulated to produce a train of 10-ns pulses with a 333-kHz repetition. CW operation was impossible to achieve at the time due to the device's inability to handle thermal issues. Of course, the pump is absorbed by all of the semiconductor materials, Si, InP, InGaAs QWs, and their InGaAsP barriers. Laser operation was achieved for various wavelengths with a threshold ranging from 1 to 7.8 mW peak power (total input power incident upon the sample). The light–light curve can be seen in Fig. 7C on a log–log scale for a cavity of 920 nm length, lasing at 1535 nm. As we can see, the response has a typical S-shape stemming from the transition from spontaneous emission to stimulated emission (Kapon, 1999). In this work, the authors perform an interesting study, which can be seen in Fig. 7D, where the experimentally obtained Q factors are compared to calculated Q factors after modeling based on photon lifetime values obtained from the rate equation fit. There is a very good matching, confirming the validity of the rate equation modeling, with the lower experimental values attesting to fabrication roughness. The initially smooth increase for shorter wavelengths is due to the fact that the modes are well confined in the "deepest" position of the gap, whereas on the longer wavelength side the sharply decreasing Q is due to the fact that the mode is now located too close to the valence band and is no longer well confined (Halioua et al., 2010).

2.1.2 L3 2D PhC Cavity With Air Bridge

The device reported in Haddadi et al. (2012) lies within the L3 family of 2D PhC cavity devices. This means that three holes are missing in the center of the triangular lattice. The target of the device was to modify and characterize the far field of the emission by employing the "band-folding" technique, in which a modulation of the radius of specific holes surrounding the cavity is introduced, allowing control over the beaming from cavity modes minimizing the off-plane leakage (Tran et al., 2009) and considerably increasing the coupling efficiency in the vertical direction (Portalupi et al., 2010; Tran et al., 2009). Although the path followed by the authors falls clearly outside the scope of this chapter, we will revisit the basic structure of the device and some of the design steps that the authors undertook, as it is one of the few optically pumped L3 PhC active nanocavities integrated on a Si substrate.

The device can be seen in Fig. 8A (folded) and Fig. 8D (unfolded). The nanocavity was InP based on a suspended membrane, having resonant modes at around 1.5–1.55 μm. The InP membrane (265 nm thick) was grown by metalorganic vapor phase epitaxy (MOVPE) and comprised four QW layers of InGaAs/InGaAsP. The QW luminescence at 300 K is centered at 1.49 μm with a spectral broadening of 86 nm. To form the air bridge, a SiO_2 sacrificial layer underneath the InP is initially bonded onto a Si substrate through a BCB layer. Consequently, wet etching is performed on the sacrificial layer lying between the InP membrane and the substrate, forming at its place the 1-μm air bridge. An example of a typical fabrication process for a nanocavity based on an air bridge can be seen in Fig. 8G.

Also in this case, the samples were pumped in the pulse mode with a 100-fs pulse duration at an 80-MHz repetition rate at 810 nm. The pumping

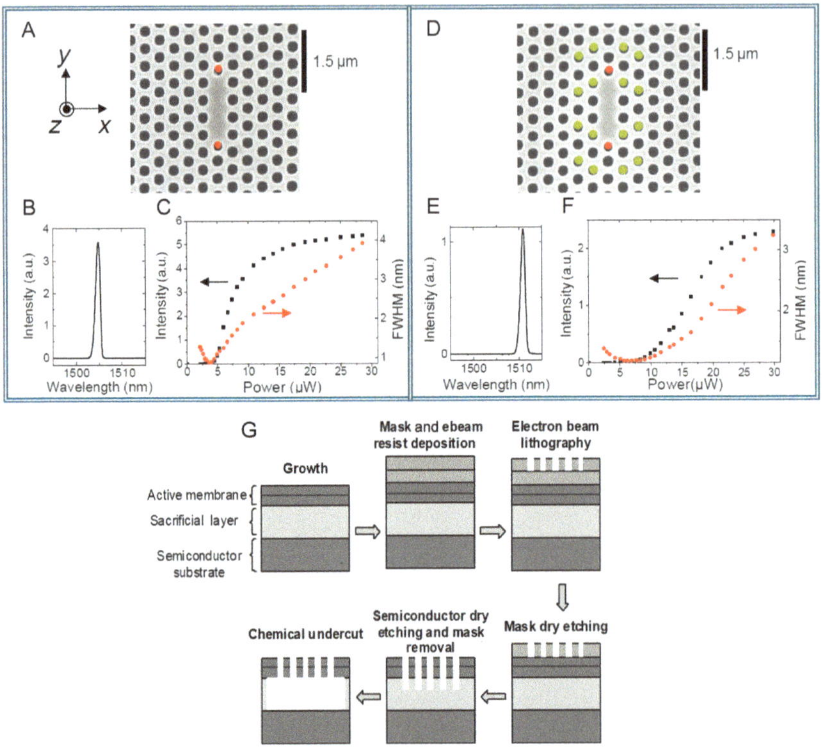

Fig. 8 SEM images of the unfolded (A) and folded (D) L3 nanocavities. *Circles filled in red* denote the shifted and shrunk-end holes to boost the *Q* factor, and those *filled in green* denote the size modulation at twice the period of the original lattice to achieve the band-folding effect. (B) and (E) are the spectral emissions of the unfolded and folded nanocavities, respectively. The output power (*black*) and spectral width (*red*) vs input power are presented in (C) for the unfolded nanocavity and (F) for the folded one, showing laser emission in both cases. (G) A typical example of the fabrication process for an air bridge-based nanocavity (Haddadi et al., 2012; Raineri et al., 2014).

power was around 10 μW. The effect that the modification of the lattice structure had in the device performance in terms of wavelength, emission intensity, and FWHM bandwidth can be seen in Fig. 8B, C, E, and F.

2.2 PhC Nanolasers on Si/SOI Waveguide Boards

2.2.1 Efficient Coupling to SOI Waveguides

The demonstrations of PhC nanolasers on plain Si wafers pointed out high-performance level of the nanolasers; however, it is imperative that the emission of the lasers is fed to the actual optical circuits. A significant work on this matter was performed in Halioua et al. (2011), where coupling is achieved between a nanowire/nanobeam PhC nanolaser and SOI waveguides. The schematic of the concept can be seen in Fig. 9A. The concept relies on two optical levels, one comprising a single-mode SOI wire waveguide and the other comprising a 1D PhC nanolaser with four InGaAsP/InGaAs emitting at 1550 nm. Between the two optical levels, there are two low-index layers of silica and BCB, which preserve the vertical optical confinement within the SOI waveguide and the PhC cavity. The exact mechanism for coupling between the two optical levels relies on the penetration of the evanescent tail of the optical modes into the other level (Haus, 1984).

The dimensions of the nanocavity are 550 nm in width and 255 nm in height and the mirrors are formed by a single row of holes drilled into the material. The hole diameter, d, and the pitch of the 1D lattice, a, are fixed to be at 177.5 and 370 nm, respectively, to obtain a large high reflectivity bandwidth around 1.55 μm. Again, a tapering region between the propagated guided mode and the mirror mode ensures the enhancement of the Q factor. The SOI level is made of a 220-nm-thick Si layer on a 2-μm SiO_2 buffer. The III–V and SOI wafers are bonded via adhesive bonding using BCB. The InP-based heterostructure was grown by MOVPE, while the SOI chip was prefabricated and prepatterned. Details on the fabrication process can be found in Halioua et al. (2011).

Beneath the cavity, the low-index layer consisting of a thin layer of BCB (80 nm) and a layer of SiO_2 provided an improvement compared to the plain BCB bonding layers as it provided better control on the InP–SOI separation but also an improvement in the thermal dissipation of the laser. It should be noted that the thermal conductivity of silica is three times greater than that of BCB.

The concept was experimentally validated and confirmed. The device was operated at RT employing optical pumping with an 800-nm laser diode focused to a 20 μm^2 spot by a 10 × microscope objective. The laser diode is modulated in order to obtain 40 ns pulses at a 300 kHz repetition rate. Measurements of the nanolaser output power were actually made for the output signal at the SOI grating couplers using SMF-28 optical fibers. As we can see in Fig. 9B, the emission curve is S-shaped, with a threshold of 17 μW for a total peak power of 2 mW. The coupling efficiency of this scheme can be easily modified to the exact percentage needed simply by adjusting the SOI waveguide width and thus its refractive index to match accordingly with that of the nanocavity level. This effect can be clearly seen

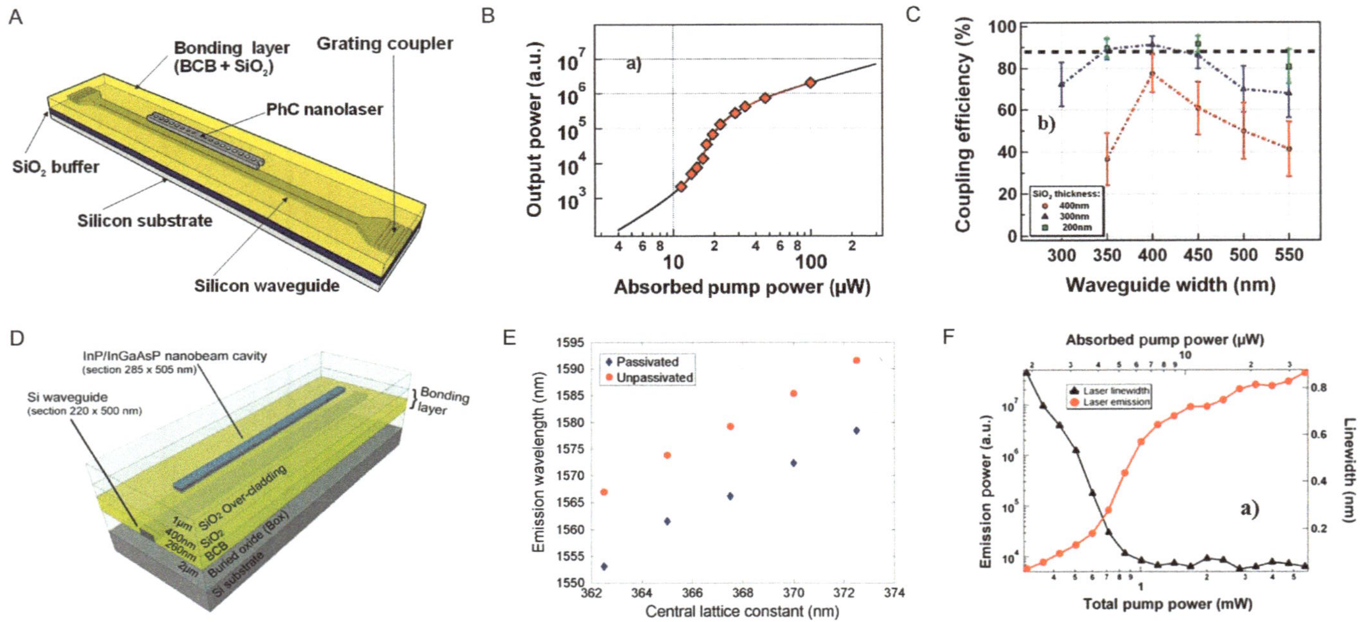

Fig. 9 (A) Schematic representation of InP-based photonic crystal wire cavity nanolaser positioned on top of an SOI strip waveguide (Halioua et al., 2011). (B) Experimentally measured intensity of the emitted light outputting the SOI waveguide as a function the absorbed pump power (Halioua et al., 2011). (C) Coupling efficiency of the emitted light into the SOI wire as a function of the structure parameters (Halioua et al., 2011). (D) Schematics of the InP-based photonic crystal wire cavity nanolaser (Crosnier et al., 2015). (E) Nanocavities emission wavelength before (*blue diamonds*) and after passivation (*red circles*) (Crosnier et al., 2015). (F) CW operation: measured emission power (*red dots*) and linewidth (*black triangles*) as a function of the pump power for a PhC nanobeam cavity with $a_0 = 365$ nm and $r = 115$ nm (Crosnier et al., 2015).

in Fig. 9C, where the coupling efficiency is adjusted at will from around 20% to more than 90% controlling the evanescent wave coupling strength through adjustment of the refractive index.

This was a greatly significant step toward bringing PhC nanolasers closer to applications, as this coupling scheme is simple, effective, and can be easily reproduced for different types of devices.

2.2.2 Elimination of Nonradiative Surface Recombination and Heat Sinking

As we have discussed previously in this chapter, nonradiative carrier recombination greatly inhibits nanolaser operation as it takes part in heating the material which leads to a decrease in the radiative efficiency and directly raises the laser threshold, making it impossible to have operation in the CW regime. In Crosnier et al. (2017), the authors describe a procedure for eliminating nonradiative surface recombination based on chemical treatment followed by encapsulation of the device.

The procedure was designed and demonstrated on the 1D PhC nanolaser coupled to SOI waveguides described earlier. The schematics of the device along with its architectural dimensions can be seen in Fig. 9D. Again, die-to-die adhesive bonding of the InP-based heterostructure on top of an SOI waveguide circuitry is performed, this time using divinylsiloxane-benzocyclobutene (DVS-BCB) diluted in mesitylene (Mes). Prior to the bonding, the III–V wafer is coated with 400-nm-thick layer of SiO_2 in order to set the distance separating the III–V and the SOI to the desired value. The DVS-BCB thickness is finally 40 nm. The III–V material is patterned using electron beam lithography with hydrogen silsesquioxane (HSQ) followed by inductively coupled plasma (ICP) etching employing $HBr/O_2/He$ chemistry. HSQ is afterward removed using fluorine-based reactive ion etching.

The surface passivation procedure in order to reduce the surface recombination was a two-step chemical treatment as follows:

1. An H_2SO_4:H_2O_2:H_2O solution at a ratio 1:8:5000 is applied for 10 s to gently etch the surface of the InGaAsP material (QWs and barriers), aiming to remove the material on the side walls damaged by the bombardment of the ions during the ICP etching.
2. The sample is bathed for 10 min in a $(NH_4)_2S$ solution at a 10% concentration, heated at 45°C. This deoxidizes the structures and fills the dangling bonds at their surface with sulfur atoms and removes the recombination centers.

Following, a significant process step involves the deposition of a 1-μm-thick silica using plasma-enhanced chemical vapor deposition so as to encapsulate the nanolasers and increase heat sinking as shown in Bazin et al. (2014a,b). This step is crucial to increase heat sinking in the levels required for CW RT operation.

As this chemical treatment actually performs a gentle etching on the surface of the material, it induces alterations in the geometric characteristics of the nanocavity. Although these alterations are very small, the optical response of PhC nanocavities is very sensitive to any change in the structure. Hence, the resonant and emission wavelengths of the nanolaser are affected, as can be seen in Fig. 9E, where a blue shift of around 15 nm can be noticed. For this measurement, the device was operated with a laser pump signal at 1180 nm with the laser diode modulated to obtain 10 ns pulses at a 250-kHz repetition rate. Calculations showed that the passivation procedure removes around 2 nm of material.

By carrying out a pump-probe experiment, the authors demonstrated a nonradiative carrier lifetime of 4.1 ns for the passivated structure, compared to the 370 ps of the unpassivated, thus showing almost an order of magnitude improvement.

Consequently, the device underwent the ultimate test of CW RT operation, employing CW optical pumping at 1.18 μm. The pump laser diode was focused down on the cavities surface to a spot size of about 5 μm using a 10× long working distance microscope objective. The emitted light is collected via the grating couplers using again SMF fibers. CW operation was successful only for the passivated samples and its measured emission power and linewidth can be seen in Fig. 9F. As seen from Fig. 9E, the emission wavelength is around 1565 nm and the S-shaped emission curve in Fig. 9F is revealing the transition from spontaneous to stimulated emission for an absorbed pump power around 8 μW (inflection point). The linewidth narrows down as the pump power is increased and takes on values under 0.1 nm. Moreover, the fact that the samples were continuously pumped for days and no degradation was observed further signifies the importance of the process.

3. ELECTRICALLY INJECTED NANOLASERS ON Si/SOI BOARDS

Despite the numerous breakthroughs achieved in the development of optically pumped PhC lasers on silicon and the demonstration of their potential in terms of power consumption, compactness, and bandwidth, having lasers with electrical injection is a prerequisite in order to apply them in most modern real-world applications. Consequently, efforts have been made for the development of such electrically powered PhC lasers. There, the main difficulty is to inject efficiently electron–hole pairs in the active medium right at the hearts of semiconductor nanocavities without inducing disastrous optical losses with metallic contact. Due to their compact sizes and the resulting tightly confined electromagnetic modes, it is an extremely challenging task.

Only four groups worldwide demonstrated so far PhC-based electrically injected nanolasers, and only they have demonstrated the possibility to integrate them onto Si or SOI. These demonstrations are the scope of the third part of this chapter.

3.1 Electrical Injection Through a Nanopillar

The first demonstration of an electrically driven PhC nanolaser was achieved by KAIST in 2004 (Park et al., 2004). The proposed structure is shown in Fig. 10.

It consists of a 2D PhC point-defect cavity made in a vertical nip InP-based heterojunction. While most of the structure is free standing, a nanopost is positioned right below the center of the cavity connecting it to a p-doped substrate. The concept relies on the exploitation of a particular cavity mode which can survive at the presence of the nanopillar. The cavity Q is calculated to be around 3500 and is said to remain in this order of magnitude when the post radius stays under $0.6a$ (with a being the PhC lattice constant). The electrical injection is ensured by contacting the n-doped semiconductor on top at the periphery of the PhC and by bringing holes

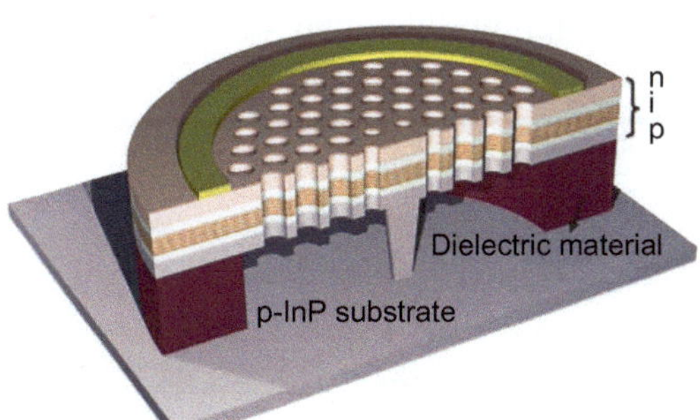

Fig. 10 Schematic representation of the first demonstration of electrically pumped PhC laser (Park et al., 2004).

through the p-doped nanopillar contacted to the InP substrate. Single-mode laser operation was demonstrated at 1519 nm at room temperature under pulsed pumping.

Certainly due to the complexity of the fabrication which involves a precise control on the semiconductor undercut for obtaining the right pillar size, it is only in 2013 that the same group adapted its idea to demonstrate an electrically driven, one-dimensional single-mode, room-temperature nanobeam laser. The features behind this selected architecture are the following:

- The reduced footprint due to the absence of a 2D PhC lattice surrounding the cavity of interest
- The easy formation of a 1D photonic bandgap, regardless of the material
- The possibility of constructing a high-Q resonator with a small number of holes, if using a high-index semiconductor material
- The possibility for nonradiative losses minimization with the proper choice of a resonant mode overlapping with the air holes

The device can be seen in Fig. 11A. The 1D cavity consists of 11 graded holes, with the cavity having a total size of $4.6 \times 0.61 \times 0.28\,\mathrm{mm}^3$, with this length of the 11 air holes agreeing reasonably with the room-temperature carrier diffusion length (Jeong et al., 2013). Again the cavity is fabricated into a vertical InP-based nip heterojunction. A submicron-sized semiconductor post is used as the p-contact. As we can also notice, the cavity is air-suspended, having contact with the p-InP layer through three micropillars. Electrons and holes injected into the n-i-p InGaAsP slab, see Fig. 11B, are recombined in the proximity of the InP post which falls within the hole diffusion. The modes originating from the air band are utilized as the region between the air holes is the only available location for the electrode, while the smaller central air holes are critical for the formation of the central underlying posts by preferential wet chemical etching which

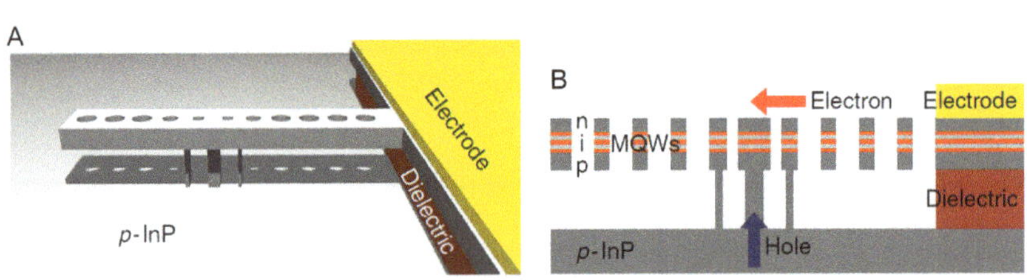

Fig. 11 (A) Schematic diagram of an electrically driven nanobeam laser. (B) Schematic diagram showing the injection of electrons and holes into the nanobeam cavity (Jeong et al., 2013).

is diffusion limited (Jeong et al., 2013). The theoretical Q factors surpass 10^3 for a structure with 11 graded air holes.

The device was operated in pulsed mode with 10-ns current pulses at a period of 1 μs at room temperature. Single-mode lasing is observed at a wavelength of 1556 nm and the measured lasing threshold current is around 65 μA. When the device was replaced with a larger three-cell nanobeam cavity by filling in three central air holes and slightly reducing the size of two air holes near the cavity, the results were improved. This time, the lasing threshold was reduced to 5 μA for 10-ns current pulses at a period of 1 μs at room temperature. As the authors state, the lower threshold of the three-cell cavity in comparison to that of the former cavity is ascribed to the higher Q factor and the smaller surface recombination losses.

This device can lead to significant conclusions for the development of electrically injected PhC nanolasers on Si. Besides the design guidelines that can be deducted from the description earlier, the poststructure of the device enables selective current injection into the central area of the nanobeam cavity with minimal tolerable additional optical scattering losses of the cavity mode. However, up to now, no solution has been proposed to integrate these lasers into a photonic circuit.

3.2 Electrical Injection Thanks to Lateral Pin Junction

Another device not integrated on Si but providing promising results was reported by Stanford University in Ellis et al. (2011) and was about a quantum-dot photonic crystal nanocavity laser in gallium arsenide where a lateral p-i-n junction was formed by ion implantation. The device can be seen in Fig. 12A, with the p-type doping region indicated in red, and the n-type region in blue. The intrinsic region is designed to be 400 nm wide in the cavity region, while it becomes wider to be 5 μm at the sides of the cavity. This is specifically intended to direct the current flow through the cavity region. For the reduction of leakage current, a trench is added to the sides of the cavity.

For the doping of the p- and n-regions, ion implantation of beryllium and silicon was performed through a mask defined by e-beam lithography for increased alignment accuracy and to avoid damaging the active region, something which is imperative for ion implantation processes. Moreover, high-temperature annealing needs to be performed to activate the dopants, raising concern for such processes to maintain compatibility with the 400°C limit of CMOS processes. The gain material for the laser comprised three layers of high-density (300 μm^2) InAs QDs.

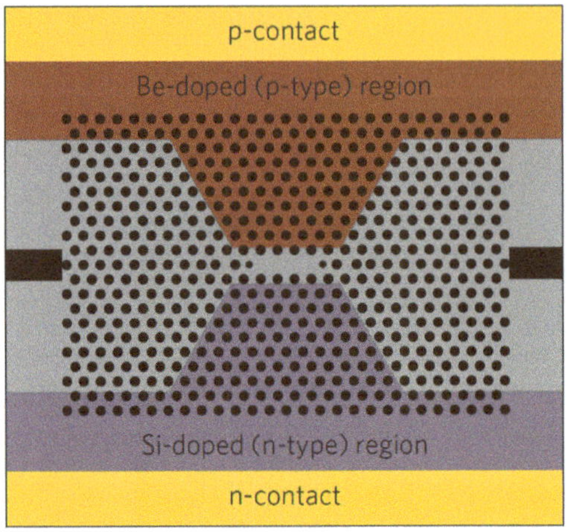

Fig. 12 Schematic of the electrically pumped photonic crystal laser. The p-type doping region is indicated in *red*, and the n-type region in *blue* (Ellis et al., 2011).

The device was able to achieve CW operation with extremely low thresholds, reaching 181 nA at 50 K and 287 nA at 150 K. However, operation at room temperature was not achieved. The low threshold achieved in this nanolaser can be attributed to a better control over the current flow, as well as the reduced surface recombination using quantum dot-based photonic crystal lasers (Englund et al., 2007a,b). The quality factor of the cavity mode at threshold was 1130. This quality factor was significantly smaller than expected from FDTD simulations, with a probable cause being the free carrier absorption or surface roughness introduced by the fabrication procedure, showing the need for further refinement of the process. Moreover, the authors noticed current leakage bypassing the cavity through the sacrificial layer and substrate, showing the need for architectural improvements of the device.

One of the most impressive efforts to achieve laser operation with a PhC structure, at room temperature and at CW mode, has been made by NTT. It revolved around the "wavelength (lambda)-scale embedded active region PhC (LEAP) laser," in which a wavelength-scale InGaAsP-based active region is embedded with an InP-based line-defect PhC waveguide (Matsuo et al., 2010, 2012). The schematic diagram of the device can be seen in Fig. 13.

An embedded active region is placed in a straight-line defect waveguide in an InP–PhC slab. The line-defect PhC waveguide has a flat surface which is important to obtain a high-Q cavity. The active region was $2.6 \times 0.3 \times 0.15\,\mu m^3$ and was placed above an air bridge of 220 nm in

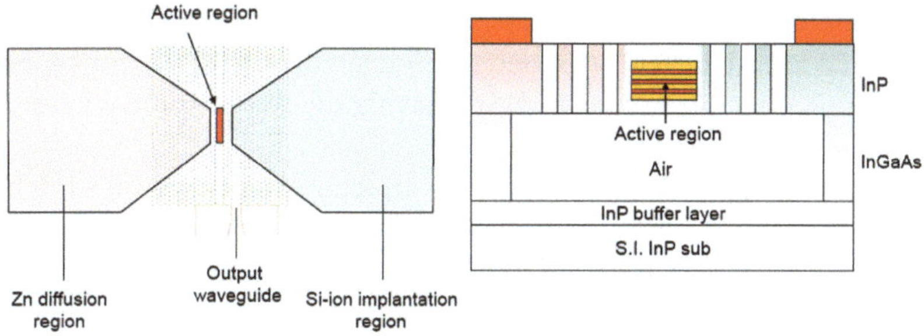

Fig. 13 Schematic diagrams of LEAP laser with lateral current injection structure: (*left*) top view; (*right*) cross-sectional view. Wavelength-scale InGaAsP-based active region is embedded with an InP-based line-defect PhC waveguide (Matsuo et al., 2012).

diameter. The aim of this structure is to reduce the thermal resistance and improve the confinement of photons and carriers in the active region, because the thermal conductivity and bandgap of the InP layer are larger than those of the InGaAsP layer (Matsuo et al., 2012).

An issue to be taken into account for electrical RT CW operation is that the current injection will inevitably increase the temperature of the active region and suppress the overall RT CW operation of the device. Consequently, actions must be taken to minimize this increase, either by driving the heat away from the active region or by minimizing the factors that lead to the temperature increase, such as the electrical resistance itself. In the latter case, actions must be taken to decrease the threshold current of the device to facilitate proper operation of the device. Such an action is the *decrease of the overall dimensions of the active region* so as to reduce the threshold current. In the case described in Matsuo et al. (2012), the active region length is decreased, again to reduce the threshold current and also *increase the Q factor to reduce the threshold current density*. As an extra means of increasing the Q factor, the output waveguide is positioned at an appropriate distance away from the cavity, as can be seen in Fig. 13. The achieved Q factor was finally 4200 and the V_{eff} was around 0.15 μm^3. The threshold current was as low as 4.8 μA, while energy efficient direct modulation was demonstrated with a bit rate of 12.5 Gbit/s and an energy cost of 14 fJ/bit for the same device in Crosnier et al. (2017).

However, the above results for the buried heterostructure LEAP laser were achieved on InP wafers, which renders cointegration with CMOS rather tricky due to regrowth step, in order to build optical interconnects for intra- or interchip communication. Consequently, efforts were made to adapt the technology and transfer the findings in the heterogeneous integration on Si platform. In Takeda et al. (2015), the fabrication process was altered so as to develop a buried heterostructure LEAP laser on Si, based on the same principle of operation but having the active material heterogeneously integrated on SOI wafer.

For the fabrication of the buried heterostructure-based laser, an i-InP layer is implanted with Si ions for n-type doping, while Zn diffusion is performed for the p-type doping. Consequently, the sample needs to be annealed at 700°C for 3′ for the activation of the dopants, followed by Zn diffusion performed at 500°C with a patterned SiO_2 layer used to define the p-type region (Matsuo et al., 2012). This is of particular importance, as these temperatures surpass by far *the CMOS process temperature limit of 400°C*, raising questions for the compatibility of this process or the requirement for radical alterations. On the other hand, one could argue that similar techniques that employ impurity implantation and diffusion are employed in silicon CMOS fabrication techniques and hence this is something that adversely affect the yield and repeatability of the process. One could notice the air-bridge structure of the device, that is an indispensable operational counterpart of the structure as it is used to maintain the quality factor of the cavity. However, this adds-up to the CMOS compatibility problems of this process.

Coupling of light is made to the silicon waveguide, which has been offsetted with respect to the cavity and PhC waveguide, contributing in the increase of the Q factor.

The main steps of the fabrication process for a buried heterostructure electrically injected laser on silicon can be seen in Fig. 14.

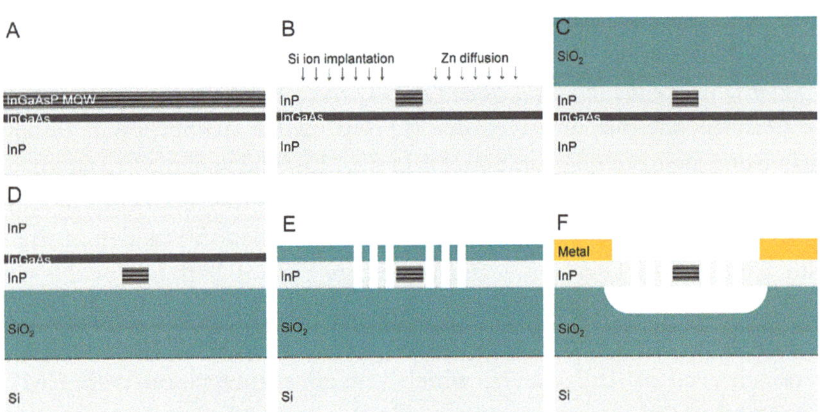

Fig. 14 Process steps for the fabrication of electrically injected buried heterostructure lasers on Si (Takeda et al., 2015).

As can be seen, the process is based on the combination of InP and Si wafers (Takeda et al., 2015). Initially, an InP substrate with an InGaAs etch-stop layer is used as the base for the MOVPE deposition to form a 6-QW buried heterostructure (Fig. 14A). e-beam lithography is used to define the active layer regions and mesa structures are formed by dry etching. After the etching, MOVPE regrowth is used for embedding the active regions. Consequently, as in Fig. 14B, the next process step involves the definition of the n-InP regions by Si ion implantation and annealing, as mentioned before, as well as the definition of the p-InP regions by the thermal diffusion of Zn, leading to the definition of the lateral p-i-n junctions. Contrary to common processes, the next step shown in Fig. 14C involves the deposition of a 2-μm-thick SiO_2 wafer on the InP. Bonding of the two wafers is performed by means of O_2 plasma-assisted bonding technique (Di Liang et al., 2011) after surface planarization by a chemical–mechanical polishing (CMP) technique (Takeda et al., 2015). Consequently, the processed InP wafer is bonded on a bulk Si substrate, Fig. 14D, using again O_2 plasma-assisted bonding. The next step involves the removal of the InP substrate and the InGaAs etch stop, as well as the definition of the PhC holes dry etching and consequent e-beam lithographic steps (Fig. 14E). The air bridge is formed by selective wet etching of the SiO_2 layer. Finally, a metal is deposited to form the contacts.

The SiO_2 layer, apart from its role as a bonding layer, is acting as an avoidance mechanism for coupling between the III–V layers and Si substrates. On the contrary, utilizing the space underneath the buried heterostructure, vertical coupling schemes can be applied, for example, by having a Si grating coupler and a Si wire waveguide integrated on Si. Nonetheless, such a scheme requires a grating coupler operating at normal incidence for optimal coupling efficiency.

The device reported in Takeda et al. (2015) comprises an active region of $2.87 \times 0.3 \times 0.15\,\mu m^3$. The InP slab is 250 nm thick, and the PhC lattice constant is 410 nm. The performance accomplished by this device includes CW operation at room temperature with a threshold current of 31 μA, showing that the goal of reducing the threshold current to achieve CW RT operation was largely on target. As can be seen in Fig. 15A, the device can achieve output powers of up to 0.27 μW—including an optical coupling loss of 10 dB—at an injected current of 200 μA, which shows that there is a road ahead for achieving powers that are enough for utilization in real system applications. A potential flaw of this approach can be noticed at Fig. 15B, where we can observe a red shift in the wavelength. This red shift is due

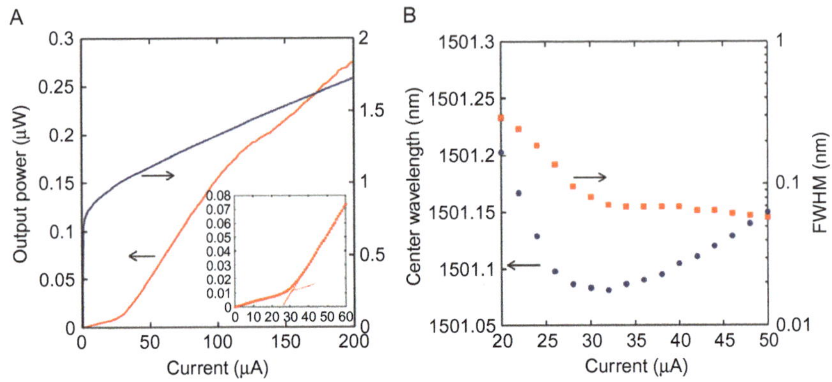

Fig. 15 (A) Measured output power and *L–I–V* curve (Takeda et al., 2015). (B) Peak wavelength and FWHM vs injected current.

to heat generation above the lasing threshold, showing that adjustments need to be made for the appropriate dissipation of heat.

3.3 Hybrid Nanorib Nanolaser Diodes

Having the aforementioned issues clearly singled out by the preceding research efforts, the device reported in Crosnier et al. (2015) was developed having the following axes at the targeted front:

- Having an efficient electrical injection scheme, while maintaining high Q factors and without disastrous optical losses induced by the metallic contacts
- Operation at CW mode and at room temperature
- Find a way to tackle nonradiative surface recombination that severely inhibits chances for CW and RT operation
- Find an efficient heat-sinking scheme
- Have a simple and efficient optical interfacing scheme with Si waveguides
- Adhere to the CMOS process requirements without regrowth

The nanolaser structure can be seen in Fig. 16A. It is a 1D PhC structure. It consists of a 450-nm-thick, 600-nm-wide, and 15-μm-long InP-based rib waveguide fully drilled with a single row of equally sized holes with a radius of 90 nm. The semiconductor base on the side of the strip is 100 nm thick and 5 μm wide and is incorporated not only to manipulate the field shape but also to facilitate the lateral placement of the p-contacts. The hole-to-hole distance gradually increases from 300 nm in the center of the structure to 330 nm toward its sides to mold the spatial profile of the resonant EM field into a Gaussian form, which keeps the radiative losses to a minimum.

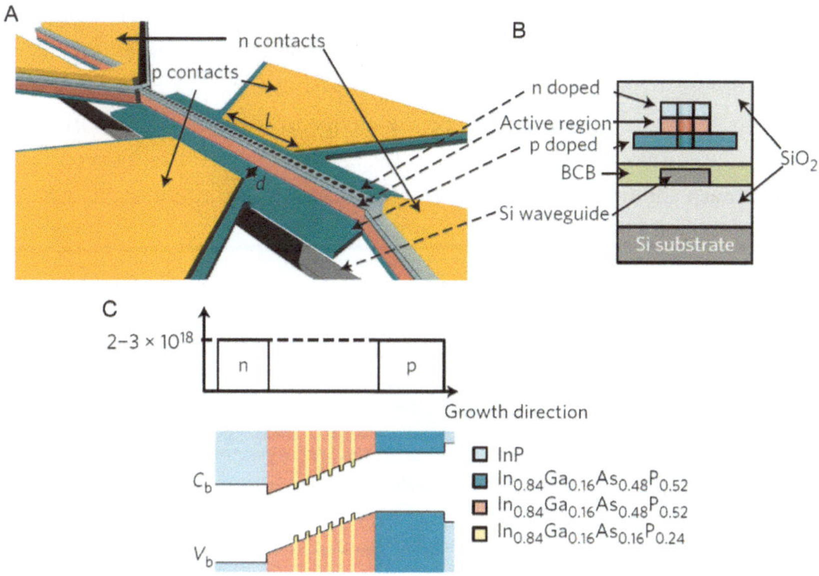

Fig. 16 (A) Schematic view of the electrically injected 1D PhC nanolaser on Si. (B) Respective cross-section showing the layer structure (Crosnier et al., 2017). (C) Doping profile of the respective heterostructure of the electrically injected 1D PhC nanolaser on Si (Crosnier et al., 2017).

This mechanism has been employed in the optically pumped structures in the past (Bazin et al., 2014a,b).

The nanolaser is placed on top of a 220 × 550-nm SOI wire waveguide. The quite convenient issue of this vertical evanescent coupling mechanism is that the percentage of coupling between the active and passive levels can be controlled at will by changing the SOI wire width, thus controlling the transverse overlap of the confined EM fields. In this design structure, the maximum coupling was achieved with the SOI wire width fixed at 550 nm. The two optical levels are separated by a transparent bilayer composed of 40 nm of BCB and 400 nm of SiO_2.

A vertical n-i-p (NIP) heterojunction is formed having the n- and p-doped layers made of InP and the i-layer of $In_{0.84}Ga_{0.16}As_{0.48}P_{0.52}$. Therefore, intraband absorption at 1.55 μm is avoided. Electrons are provided by n metallic contacts positioned on top of the structure at its far ends, whereas holes are provided by p-metallic contacts situated on top of the semiconductor base at a distance *d* from the sides of the cavity region. The exact value of this distance is extremely important and will be analyzed in this chapter. With this electrical injection configuration, the electrons travel in the n-doped layer along the semiconductor strip etched with the PhC, whereas the holes move in the p-doped layer toward the strip from its sides. Moreover, because of the large difference in the mobility of electrons (μ_e) and holes (μ_h) in the semiconductor materials ($\mu_e \approx 70\ \mu_h$), the maximum carrier radiative recombination rate occurs in the center of the cavity in front of the p-contacts.

Fig. 17A shows the spatial overlap (SOv) of the carrier radiative recombination with the EM field intensity vs the distance *d* of the metallic contacts

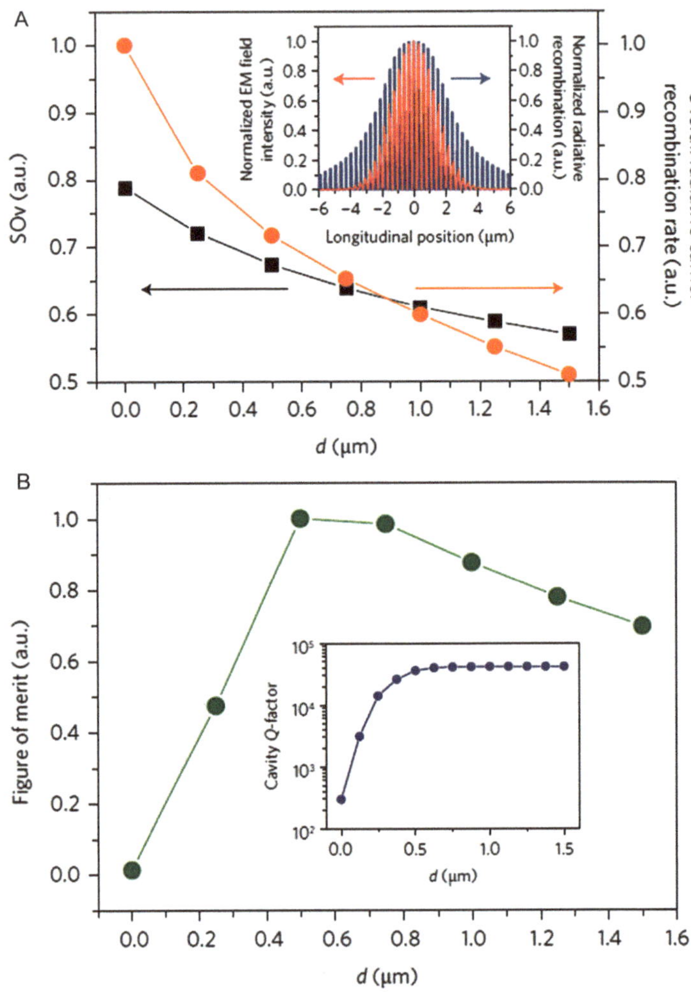

Fig. 17 (A) Simulated spatial overlap (SOv) of the electromagnetic field with the carrier recombination profile and overall carrier recombination rate. Position 0 denotes the center of the cavity (Crosnier et al., 2017). (B) Figure of merit (normalized to its maximum) calculated from the product of the overlap integral with the overall carrier radiative recombination rate (Crosnier et al., 2017).

from the sides of the cavity. Of course, the p-metallic contacts are placed in such a way so as to draw a trade-off between the electrical injection efficiency and the optical losses they induce. As can be seen, the SOv is better than 60% for $d < 1\,\mu m$ while, as can be seen in the inset, d must be larger than $0.5\,\mu m$ to maintain an intrinsic Q (Q_i) beyond 2×10^4. At the same time, the loaded quality factor (Q_L) reached a value of 2900, showing that the intrinsic cavity losses are more than one order of magnitude lower than the coupling ones, ensuring $a > 93\%$ (quasicomplete) transfer of light from the III–V PhC to the SOI waveguide.

Consequently, one could say that the aforementioned SOv value together with the intrinsic Q value and the overall recombination are crucial factors for the success of the electrical injection scheme. For this reason, the authors define a figure of merit metric to identify the structure with the potentially lowest laser current threshold. As can be seen in Fig. 17B, this corresponds to the placement of the p-metallic contacts in the region $0.5\,\mu m < d < 0.75\,\mu m$.

The fabrication of the device is based on the adhesive bonding of the III–V heterostructure that contains six strained compensated InGaAsP QWs emitting at 1.55 μm to an already patterned SOI dye (Figs. 18–22).

In this approach, the PhC nanorib is defined by two steps of electron beam lithography, that is followed by ICP etching. In this way, initially the PhC lattice is formed on top of the SOI wire and consequently the two bases on either side of the nanocavity, that have reduced height with respect to it and will allow for the deposition of the electrical p-contacts.

Consequently, the metallic contacts are defined and deposited through a lift-off procedure. For their annealing, it is important that the annealing temperature is set at 400°C, remaining within the defined limits of the process flow maximum temperature of CMOS technology.

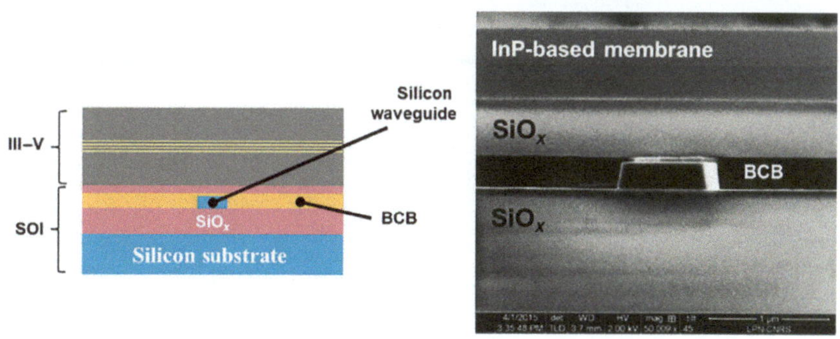

Fig. 18 BCB bonding on silicon substrate.

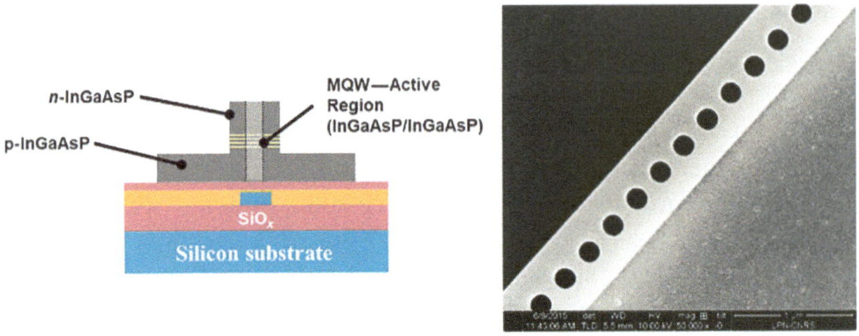

Fig. 19 Two-step ICP for photonic crystal nanorib definition.

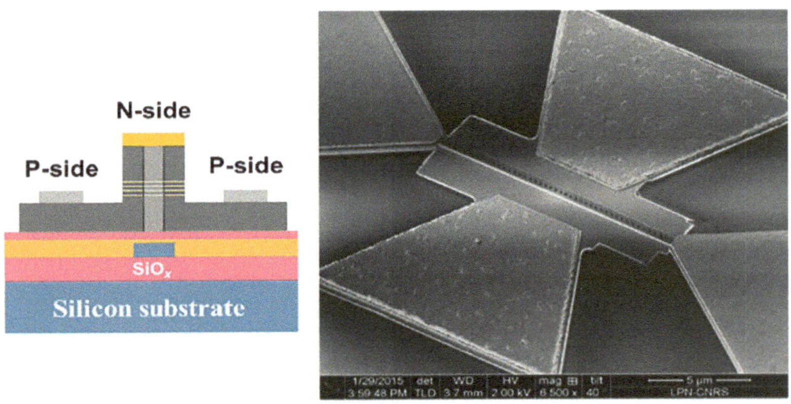

Fig. 20 Metal contacts definition through liftoff and annealing.

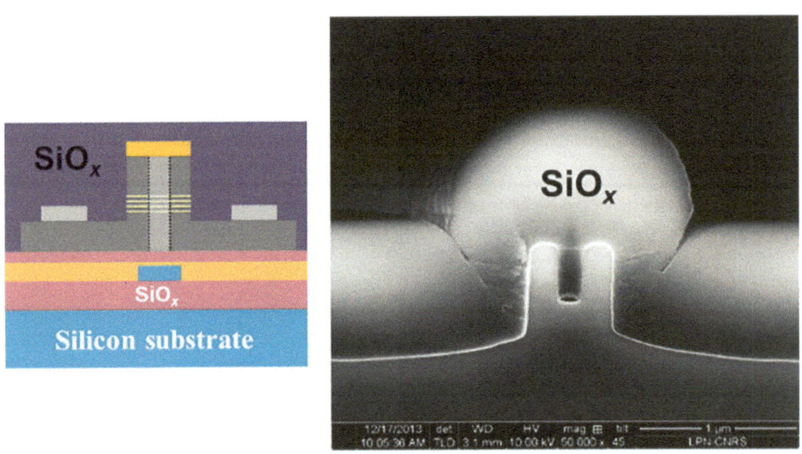

Fig. 21 Sulfur passivation and silica encapsulation.

Fig. 22 Contact pads metallization through liftoff.

The next major step in the fabrication process is the chemical passivation and consequently the encapsulation of the structure in SiO_2. The structure is passivated with a solution of sulfuric acid, which proves effective with the lengthening of the carrier lifetime from 400 ps to 2.63 ns. The encapsulation of the device in SiO_2 dramatically improves the overall heat sinking of the device. The combined action of these process steps is crucial for the CW and room-temperature operation of this type of device.

The final step of the fabrication recipe included the definition of the metal pads, the deposition of metal, and the lift-off procedure for the final contact formation. This leads to the final cross-cut image of the device, as seen below.

In total, this approach:

- methodically designs the electrical injection scheme to maintain optical properties
- exploits a simple, effective, and easily modified III–V to Si-coupling scheme
- tackles crucial issues such as nonradiative surface recombination and heat sinking

Thus it can lead to results that can really reap the benefits of PhC structures. This is evident as CW single-mode laser emission is achieved at room temperature, with the central wavelength being at 1.56 μm, having a side-mode suppression ratio of 60 dB for an electrical current of 300 μA. As can be seen from the power–current and voltage–current curves, the lasing threshold current is 100 μA and the device has a 0.75-V turn-on voltage and 1.8-kΩ serial resistance. The maximum output power coupled in the SOI wire is around 100 μW at 500 μA, a record value for an electrically injected PhC nanolaser on Si.

4. ALTERNATIVE TYPES OF PhC NANOCAVITY DEVICES AND APPLICATIONS

The development of PhC nanocavities opens new horizons for novel devices aiming at different targets but based on similar principles.

One of the most sought after concepts in photonics history has been the development of optical memories. PhC technology seemed from the start very promising to fulfill this prospect as they combine low size, low power consumption, and high-speed characteristics. As expected, the implementation of the most successful concepts in the development of PhC nanolasers could be immediately transferred to optical memory demonstrations.

One of these was the buried heterostructure device mentioned in Section 3 of this chapter, where an InGaAsP-active region ($4 \times 0.3 \times 0.15\,\mathrm{mm}^3$) is embedded within a 2D InP–PhC line defect. Thus, a high-Q nanocavity mode is formed by local mode-gap modulation for a line defect, using a combination of width and index modulations to achieve a higher Q (Nozaki et al., 2012). The InGaAsP region was buried in the InP-based PhC waveguide through etching and an overgrowth process combined with an accurate alignment technique. An image of the fabricated memory device can be seen in Fig. 23A. The device exhibited a hysteresis loop by inserting an additional pump light. The hysteresis width loops can be seen in Fig. 23C when the laser wavelength was detuned by δ from resonance, and P_{out} vs P_{in} was measured for upward/downward sweeps of the power level. Memory operation was successful and can be seen in Fig. 23D, where three pulse streams are used for the operation, with the CW bias signal being combined with the Reset signal. Although this device was not integrated on Si, it is worth noting as it is an interesting concept, while the device exhibited very low values for the bias power and switching energy—$3 \times 10^{-5}\,\mathrm{mW}$ and $2.5\,\mathrm{fJ/bit}$, respectively—with an area of $<10\,\mu\mathrm{m}^2$ for the active region. However, the very slow fall time of 7 ns renders the device incapable for the bit-level high-speed caching of computing environments and more applicable for buffering schemes in routing applications.

In the same fashion, the 1D PhC wire nanolaser, that was also described in Sections 2 and 3 of this chapter, was utilized to demonstrate memory operation (Fitsios et al., 2016). This time, injection locking was employed to exhibit bistability behavior, which can be seen in Fig. 24A. The nanolaser was pumped optically with a 1180 nm pump signal perpendicular to the direction of the PhC nanolaser output. The laser emission state toggles between a free-running mode (unlocked state) and an injection-controlled

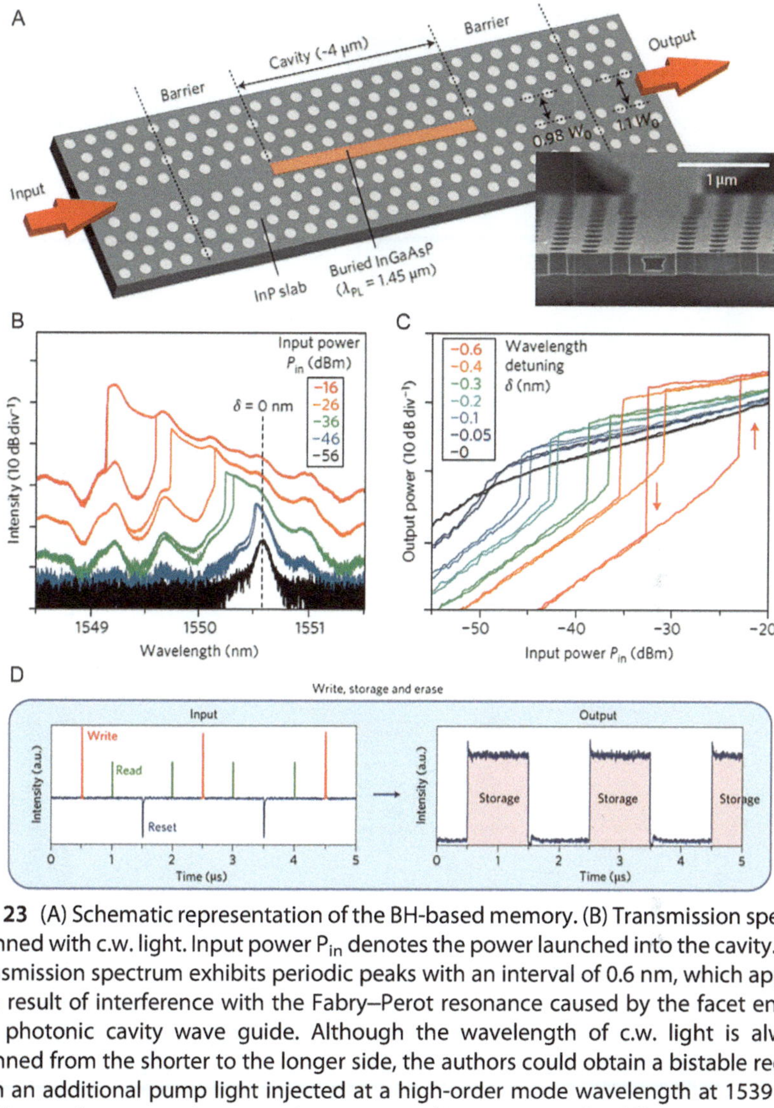

Fig. 23 (A) Schematic representation of the BH-based memory. (B) Transmission spectra scanned with c.w. light. Input power P_{in} denotes the power launched into the cavity. The transmission spectrum exhibits periodic peaks with an interval of 0.6 nm, which appear as a result of interference with the Fabry–Perot resonance caused by the facet end of the photonic cavity wave guide. Although the wavelength of c.w. light is always scanned from the shorter to the longer side, the authors could obtain a bistable region with an additional pump light injected at a high-order mode wavelength at 1539 nm. (C) Hysteresis response between the output and input powers for different amounts of wavelength detuning d of the incident c.w. light from the original resonant peak. (D) Input waveforms consisting of c.w. bias, write pulse, read pulse, and reset pulse (Nozaki et al., 2012).

mode (injection-locked state), occurring when the laser is injected with a wavelength-detuned external optical signal. When this wavelength-detuned input injection signal (control signal) is present, the laser starts emitting at the injection wavelength and not at its free-running mode wavelength when the

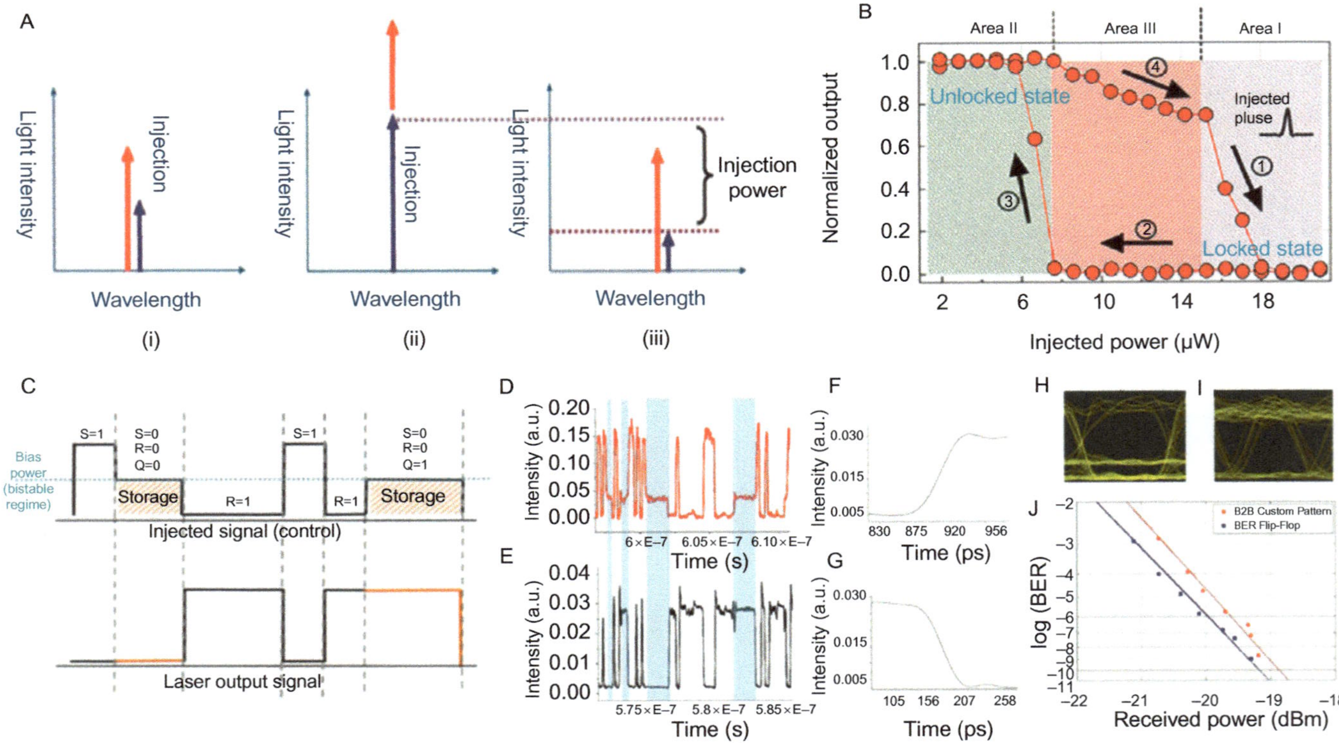

Fig. 24 (A) Process of injection locking and (B) hysteresis response of the 1D bistable PhC nanolaser. (C) Representation of the Flip-Flop operation of the 1D bistable PhC nanolaser. (D) Input pulse trace with three logical levels (logical one, bias, logical zero). (E) Rising and (F) falling time. (F) Flip-Flop output eye diagram, (G) BER curves for 1D PhC-based Flip-Flop operation at 5 Gb/s (Fitsios et al., 2016). (H) Set/reset/bias signals eye diagram, (I) memory output eye diagram, (J) BER curves for memory operation at 5 Gb/s.

control power increases above a specific threshold. While at this state, the laser can exhibit a hysteresis loop and retain its emission state even when the injected signal optical power is decreased to a certain cutoff level, returning to its free-running state only when the injected signal falls below this cutoff level.

By employing an input signal with three power states—Set and Reset which alter the state of the memory and Bias which retains the previous state, the device operated successfully at an actual data repetition rate of 5 Gb/s, which confirmed the potential of this technology for application in real computing environments. It should be noted that this device exhibits the smallest size among all PhC cavity-based optical memory implementations with a total footprint of 6.2 μm^2, raising expectations for densely integrated optical memories, while it exhibits switching times of <50 ps for switching energy of 6.4 fJ/bit, indicating operational capability in the order of 20 Gb/s.

Moreover, the performance was validated with bit-error-rate (BER) measurements at 5 Gb/s, showing error-free performance, with a negative power penalty, and raising even further the expectations for this technology in the future after these greatly encouraging results. The input and output signal streams, together with the clear eye diagrams and the BER curves, can be seen in Fig. 24H and I. With electrical injection already demonstrated for this 1D PhC nanolaser, new roads open ahead for efficient PhC optical memory configurations.

An important work that could further enhance the potential of PhC optical memory configurations was described in Hamel et al. (2015), where the authors experimentally demonstrated spontaneous mirror-symmetry breaking through a pitchfork bifurcation in a photonic molecule composed of two coupled photonic crystal nanolasers. In spontaneous breaking of inversion symmetry, a transition happens in the two localized states in the wells, which are mirror images of each other. Switching between these two states can be achieved with the injection of ultra-short pulses. The device employed was an optically pumped "photonic molecule" composed of two coupled L3 photonic crystal nanocavities with a lattice constant of $a = 425$ nm and a hole radius of $r = 0.266a$, seen in Fig. 25A. Time domain measurements of the pitchfork bifurcation can be seen in Fig. 25B. Switching between mirror images can obviously become a clear advantage for the development of optical Flip-Flops and electrical injection can improve performance and allow independent control of cavity resonances.

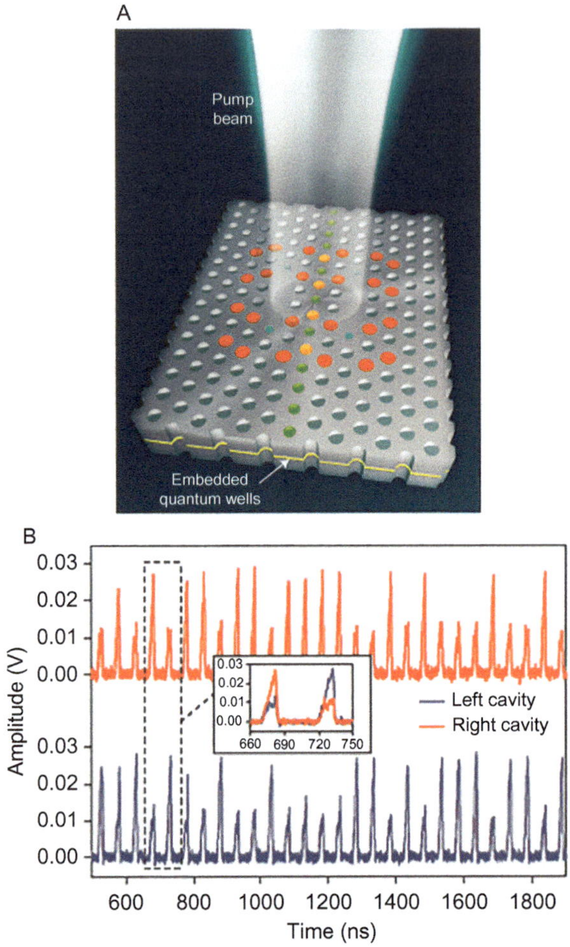

Fig. 25 (A) Three-dimensional sketch of the photonic molecule composed of two coupled L3 photonic crystal nanocavities (lattice period $a = 425\,\text{nm}$; hole radius $r = 0.266a$). Embedded QWs are represented by a *yellow line*. *Colored circles* highlight the modified holes. *Blue holes* ($r_{\text{blue}} = r - 0.06a$), shifted in the ΓK direction by $\Delta r_{\text{blue}} = 0.15a$, increase the cavity *Q* factor. *Red holes* ($r_{\text{red}} = r + 0.05a$) improve beaming of the radiated photons. *Green holes* ($r_{\text{green}} = r - 20\%$) control the coupling strength. *Orange holes* ($r_{\text{orange}} = r_{\text{red}} - 20\%$) combine *red* and *green effects*. (B) Pulse sequence of a 600-μs duration, 50-kHz output signal from both nanocavities (*blue*, L; *red*, R). The time series shows alternating *"high-red/low-blue"* and *"low-red/high-blue" states* (dead-time windows between pulses are omitted in a sequence). The zoom-in shows this alternation over two periods (Hamel et al., 2015).

REFERENCES

Akahane, Y., et al., 2003. High-Q photonic nanocavity in a two-dimensional photonic crystal. Nature 425, 944.

Asano, T., et al., 2006. Ultrahigh-Q nanocavities in two-dimensional photonic crystal slabs. IEEE J. Sel. Top. Quantum Electron. 12, 1123–1134.

Badolato, A., et al., 2005. Deterministic coupling of single quantum dots to single nanocavity modes. Science 308, 1158–1161.

Bazin, A., et al., 2014a. Design of silica encapsulated high-Q photonic crystal nanobeam cavity. J. Lightwave Technol. 32, 952–958.

Bazin, A., et al., 2014b. Thermal management in hybrid InP/silicon photonic crystal nanobeam laser. Opt. Express 22, 10570–10578.

Braive, R., et al., 2009. Transient chirp in high-speed photonic-crystal quantum-dot lasers with controlled spontaneous emission. Opt. Lett. 34, 554–556.

Coldren, L.A., et al., 2012. Diode Lasers and Photonic Integrated Circuits. Wiley Series in Microwave and Optical Engineering.

Crosnier, G., et al., 2015. Subduing surface recombination for continuous-wave operation of photonic crystal nanolasers integrated on silicon waveguides. Opt. Express 23, 27953–27959.

Crosnier, G., et al., 2017. Hybrid indium phosphide-on-silicon nanolaser diode. Nat. Photonics 11, 297–300.

Di Liang, D., et al., 2011. Uniformity study of wafer-scale InP-to-silicon hybrid integration. Appl. Phys. A Mater. Sci. Process. 103, 213–218.

Ellis, B., et al., 2011. Ultralow-threshold electrically pumped quantum-dot photonic-crystal nanocavity laser. Nat. Photonics 5, 297–300.

Englund, D., et al., 2005. Controlling the spontaneous emission rate of single quantum dots in a 2D photonic crystal. Phys. Rev. Lett. 95, 013904.

Englund, D., et al., 2007a. Efficient terahertz room temperature photonic crystal nanocavity laser. Appl. Phys. Lett. 91, 071126.

Englund, D., et al., 2007b. Low-threshold surface-passivated photonic crystal nanocavity laser. Appl. Phys. Lett. 91, 071124.

Fitsios, D., et al., 2016. Ultra-compact III–V-on-Si photonic crystal memory for flip-flop operation at 5 Gb/s. Opt. Express 24, 4270–4277.

Foresi, J.S., et al., 1997. Photonic-bandgap microcavities in optical waveguides. Nature 390, 143–145.

Fujita, M., et al., 2005. Simultaneous inhibition and redistribution of spontaneous light emission in photonic crystals. Science 308, 1296–1298.

Haddadi, S., et al., 2012. High quality beaming and efficient free-space coupling in L3 photonic crystal active nanocavities. Opt. Express 20, 18876–18886.

Halioua, Y., et al., 2009. Hybrid InP-based photonic crystal lasers on silicon on insulator wires. Appl. Phys. Lett. 95, 201119.

Halioua, Y., et al., 2010. III-V photonic crystal wire cavity laser on silicon wafer. J. Opt. Soc. Am. B 27, 2146–2150.

Halioua, Y., et al., 2011. Hybrid III-V semiconductor/silicon nanolaser. Opt. Express 19, 9221–9231.

Hamel, P., et al., 2015. Spontaneous mirror-symmetry breaking in coupled photonic-crystal nanolasers. Nat. Photonics 9, 311–315.

Haus, H.A., 1984. Waves and Fields in Optoelectronics. Prentice-Hall.

Hostein, R., et al., 2010. Demonstration of coherent emission from high-Q photonic crystal nanolasers at room temperature. Opt. Lett. 35, 1154–1156.

Huang, J., et al., 2001. Room temperature, continuous-wave coupled-cavity InAsP/InP photonic crystal laser with enhanced far-field emission directionality. Appl. Phys. Lett. 99, 091110–091113.

Jeong, K.-Y., et al., 2013. Electrically driven nanobeam laser. Nat. Commun. 4, 2822.
Joannopoulos, J.D., et al., 2008. Photonic Crystals: Molding the Flow of Light. Princeton University, second ed.
John, S., 1987. Strong localization of photons in certain disordered dielectric superlattices. Phys. Rev. Lett. 58, 2486.
Johnson, S.G., 2003. Introduction to Photonic Crystals: Bloch's Theorem, Band Diagrams, and Gaps (but No Defects). MIT tutorial. ab-initio.mit.edu/photons/tutorial/photonic-intro.pdf.
Kapon, E., 1999. Semiconductor Lasers. Academic Press.
Kounoike, K., et al., 2005. Investigation of spontaneous emission from quantum dots embedded in two-dimensional photonic-crystal slab. Electron. Lett. 41, 1402–1403.
Koyama, F., et al., 1991. Intensity noise and polarization stability of GaAlAs-GaAs surface emitting lasers. IEEE J. Quantum Electron. 27, 1410–1416.
Kuksenkov, D.V., et al., 1997. Spontaneous emission factor in oxide confined vertical-cavity lasers. Appl. Phys. Lett. 70, 13–15.
Kuramochi, E., et al., 2006. Ultrahigh-Q photonic crystal nanocavities realized by the local width modulation of a e defect. Appl. Phys. Lett. 88, 041112.
Lalanne, P., Hugonin, J.P., 2003. Bloch-wave engineering for high-Q, small-V microcavities. IEEE J. Quantum Electron. 39, 1430–1439.
Laurent, S., et al., 2005. Indistinguishable single photons from a single quantum dot in a two-dimensional photonic crystal cavity. Appl. Phys. Lett. 87, 163107.
Matsuo, S., et al., 2010. High-speed ultracompact buried heterostructure photonic-crystal laser with 13 fJ of energy consumed per bit transmitted. Nat. Photonics 4 (9), 648–654.
Matsuo, S., et al., 2012. Room-temperature continuous-wave operation of lateral current injection wavelength-scale embedded active-region photonic-crystal laser. Opt. Express 20, 3773–3780.
Nomura, M., et al., 2006. Room temperature continuous-wave lasing in photonic crystal nanocavity. Opt. Express 14, 6308–6315.
Nomura, M., et al., 2007. Temporal coherence of a photonic crystal nanocavity laser with high spontaneous emission coupling factor. Phys. Rev. B 75, 195313.
Nozaki, K., et al., 2012. Ultralow-power all-optical RAM based on nanocavities. Nat. Photonics 6, 248–252.
Park, H.-G., et al., 2004. Electrically driven single-cell photonic crystal laser. Science 305, 1444–1447.
Portalupi, S.L., et al., 2010. Planar photonic crystal cavities with far-field optimization for high coupling efficiency and quality factor. Opt. Express 18, 16064–16073.
Purcell, E.M., et al., 1946. Resonance absorption by nuclear magnetic moments in a solid. Phys. Rev. 69, 37.
Quan, Q., Loncar, M., 2011. Deterministic design of wavelength scale, ultra-high Q photonic crystal nanobeam cavities. Opt. Express 19, 18529–18542.
Raineri, F., et al., 2014. Optically pumped semiconductor photonic crystal lasers. In: De La Rue, R.M., Siyuan, Y., Lourtioz, J. (Eds.), Compact Semiconductor Lasers. Wiley. https://doi.org/10.1002/9783527655342.ch2.
Romeira, B., Fiore, A., 2018. Purcell effect in the stimulated and spontaneous emission rates of nanoscale semiconductor lasers. IEEE J. Quantum Electron. 54 (2), 1–12.
Sakoda, K., 2005. Optical Properties of Photonic Crystals. 2005 Springer-Verlag.
Sauvan, C., et al., 2005. Modal-reflectivity enhancement by geometry tuning in Photonic Crystal microcavities. Opt. Express 13, 245–255.
Song, B.-S., et al., 2005. Ultra-high-Q photonic double heterostructure nanocavity. Nat. Mater. 4, 207–210.
Strauf, S., et al., 2006. Self-tuned quantum dot gain in photonic crystal lasers. Phys. Rev. Lett. 96, 127404.

Takeda, K., et al., 2015. Heterogeneously integrated photonic-crystal lasers on silicon for on/off chip optical interconnects. Opt. Express 23, 702–708.

Takeda, K., et al., 2017. In: Si nanowire waveguide coupled current-driven photonic-crystal lasers. 2017 Conference on Lasers and Electro-Optics Europe European Quantum Electronics Conference (CLEO/Europe-EQEC).

Tanaka, Y., et al., 2008. Design of photonic crystal nanocavity with Q factor of $\sim 10^9$. J. Lightwave Technol. 26, 1532–1539.

Tran, N.-V.-Q., et al., 2009. Directive emission from high-Q photonic crystal cavities through band folding. Phys. Rev. B 79, 041101(R).

Velha, P., et al., 2006. Ultra-high-reflectivity photonic-bandgap mirrors in a ridge SOI waveguide. N. J. Phys. 8 (9), 204.

Vuckovic, J., et al., 2002. Optimization of the Q factor in photonic crystal microcavities. IEEE J. Quantum Electron. 38, 850–856.

Xu, Y., et al., 2000. Quantum analysis and the classical analysis of spontaneous emission in a microcavity. Phys. Rev. A 61, 033807.

Yablonovitch, E., 1987. Inhibited spontaneous emission in solid state physics and electronics. Phys. Rev. Lett. 58, 2059.

Zhang, Y., et al., 2010. Photonic crystal nanobeam lasers. Appl. Phys. Lett. 97, 051104.

FURTHER READING

Ahn, B.H., et al., 2010. One-dimensional parabolic-beam photonic crystal laser. Opt. Express 18, 5654–5660.

Ellis, B., et al., 2010. Electrically pumped photonic crystal nanocavity light sources using a laterally doped p-i-n junction. Appl. Phys. Lett. 96 (18), 181103.

Long, C.M., et al., 2009. Modified spontaneous emission from laterally injected photonic crystal emitter. Electron. Lett. 45 (4), 227–228.

Matsuo, S., et al., 2013. Ultralow operating energy electrically driven photonic crystal lasers. IEEE J. Sel. Top. Quantum Electron. 19 (4), 4900311.

CHAPTER SIX

Heterogeneous Integration of III–V Lasers on Si by Bonding

Michael L. Davenport[1], Minh A. Tran, Tin Komljenovic, John E. Bowers
Department of Electrical and Computer Engineering, University of California, Santa Barbara, CA, United States
[1]Corresponding author: e-mail address: davenport000@gmail.com

Contents

1. HETEROGENEOUS INTEGRATION BY BONDING

Integrated photonics bring a promise of significant size, weight, power, and cost (SWaP-C) reduction, as well as significant performance and reliability improvements due to tight integration of components and reduced coupling losses. Many platforms are used to realize integrated photonics, but for efficient electrically pumped on-chip light generation, a direct bandgap semiconductor is a necessity. Historically, the first integrated photonic circuits with lasers were realized on GaAs and InP substrates (Coldren et al., 2012). More recently, optical components are being integrated on silicon (Si) substrates, whose processing was perfected by the remarkable

Semiconductors and Semimetals, Volume 99
ISSN 0080-8784
https://doi.org/10.1016/bs.semsem.2018.07.001

investment from the microelectronics industry. The maturity of Si processing, including the uniformity and yield, is unrivaled and is an important motivation for realizing photonics in Si. Photonic circuits integrated on Si substrates using CMOS process flows also allow easier 2.5D or 3D integration with electronics. In the past 15 years, a whole range of photonic components have been demonstrated, mostly in the silicon-on-insulator (SOI) platform, which is especially suited for standard communication bands at 1.31 μm (datacom) and 1.55 μm (telecom), as Si and its oxide SiO_2 are both transparent and form high-index contrast, high-confinement waveguides ideally suited for medium to high-integration PICs with low propagation losses. However, electrically pumped efficient optical sources on silicon remain a challenge due to its indirect bandgap. There have been some demonstrations with strained Ge, but the performance is very far from practical products, and threshold currents are typically 1000 × higher than what is demonstrated with III–V on Si (Camacho-Aguilera et al., 2012). A number of current Si photonic PICs use hybrid approaches in which Si PIC is coupled to III/V PIC that provides gain and lasing (Kobayashi et al., 2015). Such an approach requires precise alignment and arguably has scaling limitations that limit the integration of III–V materials to one or two areas on the PIC.

An alternative approach, most commonly called heterogeneous integration (but also historically hybrid integration which breeds some confusion), bonds pieces of III/V materials on patterned Si wafer and then processes them together using standard lithography tools. First demonstrated by UCSB in 2006 (Fang et al., 2006b), such an approach is currently widely used by academia, research organizations, and industry with commercial products for data center market being shipped in volume (Duan et al., 2015; Heck et al., 2013; Koch et al., 2013; Komljenovic et al., 2016; Liang et al., 2016; Roelkens et al., 2015). The advantage of this approach is that the alignment tolerances are significantly reduced as devices are defined using lithography alignment marks after bonding of larger III/V pieces, which allows for efficient volume scaling.

Heterogeneous silicon photonics works to provide the best of both worlds: high-performance devices from optimized III/V materials and superior passive and waveguide technology of the SOI platform. In the last 10 years, many high-performing components realized in the heterogeneous silicon platform have been demonstrated, some of which outperform native III/V components. The extraordinary performance can be attributed to at least two factors: reduced propagation losses and added flexibility in controlling the overlap between the optical field and the carriers. Low noise lasers, either narrow-linewidth single-frequency or short pulse mode-locked lasers, are definitely one of key strengths of the platform, and the remainder of the chapter will specifically focus on them. Here, we just mention record high-bandwidth high-power detectors with 12 dBm at 40 GHz (Xie et al., 2015) and −2 dBm at 70 GHz with very low dark currents (Wang et al., 2018).

The heterogeneous platform has further been extended, in recent years, beyond near-infrared telecommunication wavelengths to the mid-infrared (2–20 μm) regime using a number of waveguide platforms (Si, SiO_2, SiN, Ge…) and laser technologies such as InP Type-I and Type-II diodes, GaSb Type-I diodes, interband cascade laser (ICL) structures, and InP, GaAs, and InAs quantum cascade laser structures but this lies outside the scope of this chapter (see, e.g., Spott et al., 2017).

1.1 Fabrication

The general method for fabrication of a heterogeneous silicon/III–V device is to first form the silicon photonic devices, and then attach unprocessed dies or full wafers of the III–V material. Attachment of dies is more common, as III–V wafers are much smaller than silicon wafers, and typically the III–V is not needed over the entire die, allowing efficient use of the more expensive III–V material. The most common combination is silicon and InP, due to the well-established capability of this III–V material for emission in the C- and O bands. Bonding of GaAs substrates, for example, quantum dot lasers with O-band emission (Kurczveil et al., 2016) is also possible, along with the GaSb ICL lasers mentioned earlier. Selection of the III–V substrate must be considered carefully, as most bonding methods require an anneal after bonding. Because of this, the two substrates will be subject to stress due to thermal expansion coefficient mismatch, which can cause failure to bond or reduction in the electrooptical efficiency of the material due to the formation of defects. The coefficient of thermal expansion of the III–V substrate relative to silicon must be considered (Moutanabbir and Gösele, 2010).

Devices are typically formed by attaching the III–V material over a silicon waveguide and fashioning a III–V ridge that operates as a tandem or separate waveguide. Normally there is some kind of coupler at either end of the III–V ridge to interface with subsequent passive silicon waveguides. A schematic example of this is shown in Fig. 1.

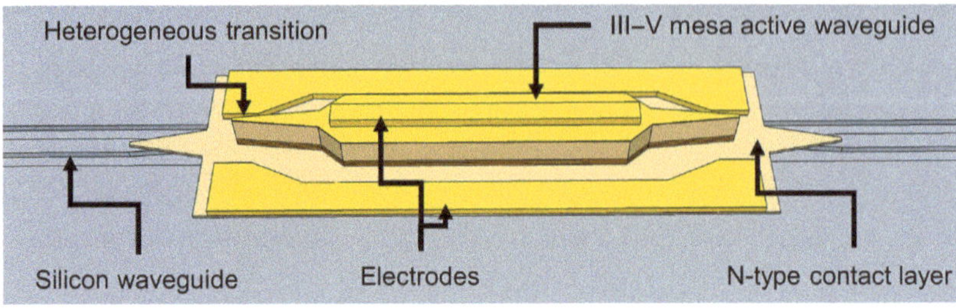

Fig. 1 Schematic diagram of a III–V ridge fabricated using heterogeneous integration.

1.1.1 Process Flow

In most cases, the III–V waveguide forms a twin waveguide system with the silicon. A hybridized optical mode will be formed, with varying degree of overlap between the silicon and III–V depending on the geometry of the cross section. The earliest demonstrations directly bonded the III–V in extremely close proximity to the silicon waveguide (Fang et al., 2006a), creating a quasi-single-mode hybridized waveguide, while subsequent versions of the technology introduced a planarized interlayer that increases the separation between the waveguides, allowing the mode to be transferred fully from one waveguide to the other (Keyvaninia et al., 2013b). The interlayer bonded devices have the advantage of higher confinement factor, while directly bonded devices have better thermal properties and therefore higher maximum output power (Davenport et al., 2016).

The fabrication process flow of an example of a directly bonded silicon/InP device is shown in Fig. 2. The first step is the entire silicon device processing, shown in Fig. 2A. This may include implantation for doping of modulators, regrowth of germanium, metallization, and via formation, but at the minimum requires an etch to define the silicon waveguide. Fig. 2B shows the device after bonding of the III–V material and removal of the substrate. In the case of InP, this is quite straightforward: including an InGaAs etch-stop layer allows the substrate to be removed with a wet etch. In most cases, this InGaAs layer is already included in the device for the p-type contact. Then, in Fig. 2C, the III–V mesa is defined. This usually consists of a series of etches, first to etch through the InGaAs, if present, then to etch the InP cladding, which is typically very thick (1.5 μm, for example), then the active region, which may be AlGaInAs or InGaAsP in most cases, then finally to etch the bottom InP cladding. Dry etching is usually preferred for the InP etch and can be combined with the initial InGaAs etch with most dry etch chemistries. Etching of the active region is normally

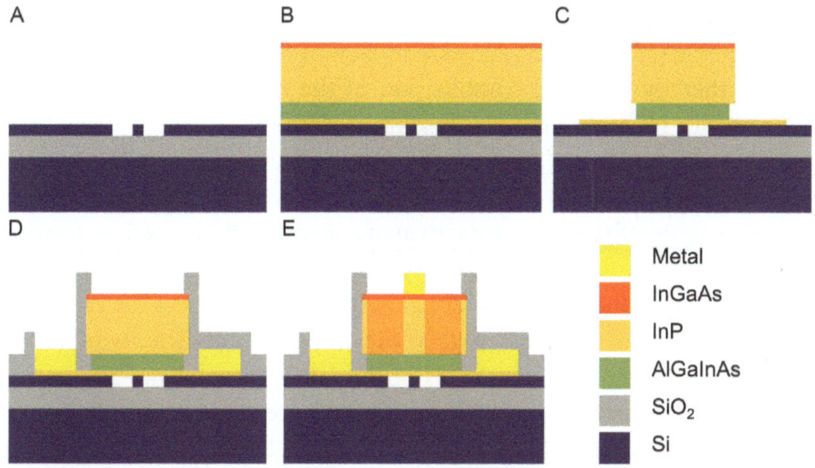

Fig. 2 The process flow of a heterogeneous silicon/InP laser waveguide: (A) shows the processed SOI waveguide, (B) is after the bonding and removal of the InP substrate, (C) is after the III–V mesa etch, (D) shows the *bottom* n-metal contact and via oxide formation, and (E) shows the device after the *top* p-metal contact and isolation implant.

conducted with a wet etch to stop on the bottom cladding. The bottom cladding can be removed with a dry etch, specifically a methane–hydrogen chemistry, to avoid damaging the underlying silicon. In the case of an interlayer bonded device, the bottom cladding can also be wet etched away; this is inadvisable in a directly bonded device, as the wet etchant can penetrate into the trenches formed by the silicon waveguide and damage the device from underneath.

After this point, the fabrication is similar to a typical III–V ridge laser device. Deposition of the n-contact metal electrodes, and the deposition and etching of a dielectric via and cladding layer, is shown in Fig. 2D. The p-contact metal electrode is then deposited, and in the case of wide-mesa designs (see Section 2.1), used as a mask to block hydrogen implantation of the mesa to form the current channel, as shown in Fig. 2E. Depending on the design, the device may then require additional metal interconnect and pad layers for packaging (not shown).

1.2 Covalent Molecular Direct Bonding

Bonded heterogeneous materials are typically either bonded directly to the silicon or bonded to a planarizing interlayer, such as the polymer divinylsiloxane-bis-benzocyclobutene (DVS-BCB) or polished silicon dioxide. The earliest examples of bonded devices used a hydrophobic process, otherwise known as wafer fusion, to integrate the heterogeneous materials (Hawkins et al., 1996). This allows conduction through the bonding interface, and thereby electrical combination of the materials as well as optical. However, this process requires annealing above 600°C, which degrades the material quality and limits application of the technique.

Alternatively, a hydrophilic bonding process may be used, where a passivating layer of typically an activated native oxide is used to form the bonding layer (Liang et al., 2011b). Hydrophilic processes can be conducted at much lower temperatures, as low as 300°C, which allows for integration of a wider range of materials with more disparate coefficients of thermal expansion. Bonding very thin layers of the III–V material can also increase the robustness of the device to high temperature, which can allow high-temperature processes at 600°C, such as regrowth, and maintain high enough material integrity to produce low threshold lasers (Matsuo et al., 2014).

The majority of heterogeneous silicon/InP devices demonstrated so far have used a hydrophilic bonding process, typically the oxygen plasma activation process, where both the silicon substrate and the III–V chip are exposed to an oxygen plasma discharge before being placed into physical contact and annealed. Annealing is often conducted under pressure. Direct bonding of InP to silicon with oxygen plasma activation, for example, will produce successful bonds with 200 kPa pressure and 300°C annealing for 60 min. Lower temperatures can produce strong bonds if combined with longer annealing times.

Beyond the oxygen plasma activation, little surface preparation is needed. The III–V chip, if bonding dies instead of wafers, should be coated with a protective layer like photoresist before cleaving to prevent crumbs of the fragile III–V from readhering to the wafer surface. This resist can be stripped directly before bonding. The silicon wafer, which has been processed, should be cleaned carefully. Cleaning with piranha solution (3:1 H_2SO_4:H_2O_2) and then a mechanical method such as ultrasonic swabbing should be conducted until the bonding area is free of particulates.

For most hydrophilic bonding processes, surface water trapped at the bonding interface can result in voids in thin films. The addition of a channel or reservoir on the silicon chip can allow these trapped gasses to diffuse away and prevent the formation of voids. Fig. 3 shows three different patterns and the impact that they have on void formation on an SOI wafer with a 150-nm InP layer transferred to the SOI surface with oxygen plasma-activated direct bonding. The lower left quadrant of the image shows an unpatterned section of the chip; many voids can be seen. The upper left

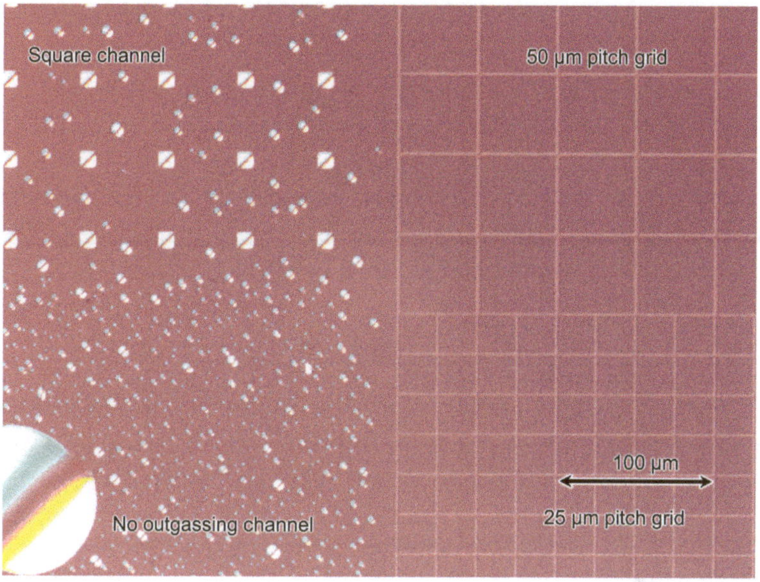

Fig. 3 An optical micrograph of an SOI chip with a 150-nm film of InP bonded over it. The four quadrants of the image have different patterns etched into them 500 nm deep, all the way to the buried oxide. The *upper left quadrant* shows a *square vertical channel*, the two *right quadrants* show grid outgassing channels, and the *lower left quadrant* is unpatterned.

quadrant has 10 µm by 10 µm square reservoirs etched into it and shows a significant reduction in the size of the voids compared to the unpatterned section. The right side of the chip has a continuous grid pattern etched into it, with two different pitches on the top and the bottom. These have the same surface area fill factor as the square reservoirs but demonstrate a complete elimination of the void patterns. It should be noted that while the continuous grid pattern could act as a channel for the gas to escape, they are effective even if the entire pattern is covered with InP.

The yield of the bonding process is an important point of discussion. In theory, the bonding process will have 100% yield, provided that the surfaces are perfectly flat, smooth, and clean. However, surface roughness (above 1 nm RMS) can degrade the bonding yield, and roughness above 3 nm RMS may prevent bonding entirely. Prime InP epitaxial wafers typically have roughness below 0.5 nm RMS, and SOI wafers are typically below 0.2 nm RMS, so for unimpaired material this is not normally an issue. Particles that collect on the wafer surface will also impair bonding yield, particularly hard noncompressible particles. A particle will often create a hole in the bonded film as much as 10 times the size of the particle. Also, the pattern on the silicon wafer must be considered. High-fill-factor etches, such as dense waveguides, reduce the amount of surface area for bonding, and the overall strength of the bond can decrease.

Provided that the surface quality is maintained, and the wafers are clean of particulate contamination, extremely high bonding yield is possible. In Liang et al. (2011b), a full 150mm InP wafer was bonded to a 150-mm SOI wafer, with only a single significant bonding defect.

2. HETEROGENEOUS LASER ARCHITECTURES

The heterogeneous silicon photonics platform opens up a new possibility in improving the noise performance of lasers by providing a mechanism to separate the photon resonator and the highly absorbing active medium by both the transverse and the longitudinal control of confinement (Davenport et al., 2016; Santis et al., 2014). The transverse confinement can be controlled by optimizing the number of quantum well (QW) layers, design of the separate-confinement heterostructure (SCH) layers, and the underlying Si waveguides. The transverse confinement can furthermore continuously change by adjusting the width of the Si waveguide, which is lithographically defined and highly controllable. The longitudinal confinement is controlled by the ratio between the active and the passive section lengths, where the transition is crucial for good performance. We address that transition in Section 2.2. In a traditional III–V semiconductor laser, there is an inevitable compromise as GaAs/InP waveguides are inherently lossier than Si waveguides. The heterogeneous platform breaks the compromise and allows photon storage in silicon and using III–V for gain resulting in record performance.

2.1 Amplifier Waveguide

The amplifier waveguide section is central to the performance of any laser design, since it consists of both the electrical and optical confinement structures, as well as the electrical contact and current injection structures. Highly efficient lasers and amplifiers depend on optimization of these structures. The situation is complicated by the fact that the electrical properties, which are important for low operating voltage and low self-heating, are often in opposition to the optical properties, specifically the waveguide loss, as higher conductivity epitaxial layers have consequently higher optical absorption (Casey and Carter, 1984; Gonzalez et al., 2009).

For most applications, a single transverse optical mode is required. This is because the intracavity filter used to produce a single-wavelength laser will typically produce a second filter response peak at a different wavelength for higher order transverse modes, which have lower effective indices than the fundamental mode. In addition, multiwavelength lasers, such as comb lasers and mode-locked lasers, will have a mode-beating envelope imposed on their output spectra as a result of mixing between transverse modes, which will limit their wavelength bandwidth. Coupling to free-space or single-mode fibers will also be impaired by multimode lasers, as the higher order modes will not couple efficiently into a single-mode fiber or simple ball lenses.

Waveguides normally must be very small to be *fundamentally* single mode (where only one transverse mode is supported), on the order of a few wavelengths inside the waveguide material. Monolithic indium phosphide waveguides must typically be less than 2 μm wide. Unfortunately, this can be detrimental to the electrical and thermal properties of the device, as the size of the electrical contact on the top of the waveguide is usually the p-type contact that determines the bulk of the diode forward resistance and shrinking this contact will increase this resistance. The thermal conductivity of narrow ridges can also be problematic. For these reasons, laser waveguides are often designed to be *quasi*-single mode, where only one of the transverse modes has positive net gain. This can be accomplished by increasing the propagation loss of the higher order mode by radiation loss, increased interaction with the sidewall, or reducing the confinement factor of the mode with the amplifying active region. The confinement factor in the active region, optical propagation loss, and the thermal and electrical properties are all affected by the transverse dimensions of the waveguide and must be optimized carefully to ensure high performance.

Another consideration at this stage of the design is the impact of the design of the amplifier waveguide on the heterogeneous transition. Integration with other components—such as modulators and passive devices—requires a transition from the heterogeneous region into a passive, typically purely silicon, waveguide. These transitions are often a point of impairment in heterogeneous silicon/III–V lasers, and so ensuring that a low loss, low reflection, and low parasitic higher order mode emission transition can be achieved from the amplifier waveguide section into a passive silicon waveguide is important for the ultimate performance of the device.

2.1.1 Cross Section Design

The first consideration that must be addressed is the optical mode profile in the transverse cross section of the amplifier waveguide. Because the silicon waveguide must support a tightly confined optical mode by itself for purposes of integrating compact and tightly bend passive devices and modulators, the addition of the InP layers will generally create complications to the optical mode. As mentioned earlier, it is crucial that only a single mode—preferably the fundamental lowest order mode—has positive net gain in the amplifier waveguide.

The addition of the III–V layers creates an unavoidable higher order transverse mode. In the case of directly bonded lasers, where conduction through the bonded interface is impossible, the lowest layer in the III–V structure is normally InP, while the quantum well and SCH layers are a quaternary with higher index of refraction. This creates a second waveguide core, forming a twin waveguide system much like a directional coupler. Due to this, the cross section will guide these two modes even if the waveguide dimensions are reduced to extreme small sizes, which is detrimental to the thermal, electrical, and optical properties.

Fig. 4A and B shows the transverse electric field of the optical mode profile of a waveguide with a narrow III–V ridge and single-mode (without the III–V) silicon waveguide. The contours indicate the mode-field amplitude, and the gray scale backdrop represents the index of refraction. The lower shallow rib waveguide is the silicon, while the tall upper waveguide is the III–V. Note in Fig. 4A, the symmetric mode where both lobes have the same sign and there is a minimum in the field centered over the thin InP contact layer that separates the two waveguides. In Fig. 4B, the antisymmetric mode has opposite sign of the field amplitude with a zero-crossing over the thin InP contact.

Excitation of the antisymmetric mode is extremely undesirable, as it will introduce a phase-matching problem at the output of the amplifier waveguide section, be it to a passive silicon waveguide or a single-mode fiber. In addition, the two modes could beat along the longitudinal dimension of the amplifier, creating periodic nulls and peaks in the optical power concentration, which is undesirable for single-frequency lasers due to spatial hole burning. The two modes also have slightly different effective indices, which will introduce wavelength-dependent modulation on the laser output spectrum.

Elimination of the antisymmetric mode can be achieved by widening the III–V mesa. While this is counterintuitive, the antisymmetric mode,

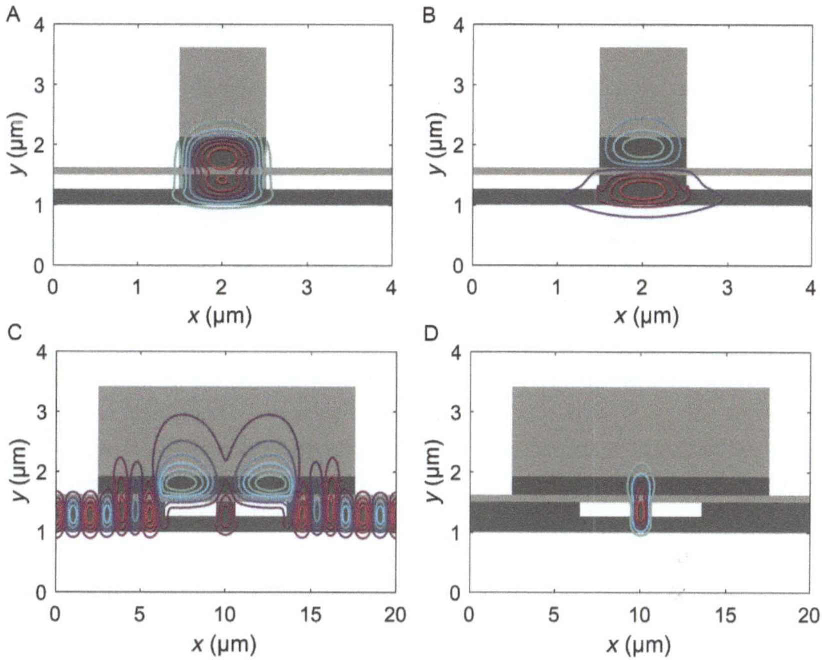

Fig. 4 Simulated transverse mode profile of the transverse electric field of a directly bonded heterogeneous amplifier waveguide. The index profile is plotted as gray scale, with darker indicating higher index of refraction. The electric field is plotted as a contour. (A) The fundamental mode with the highest effective index of a twin waveguide with a narrow silicon rib and narrow III–V ridge. (B) The next higher order mode, which is in this case the antisymmetric mode. (C) The antisymmetric mode after the introduction of the wide III–V mesa and silicon slab. (D) The fundamental symmetric mode after the introduction of a wide III–V mesa and silicon slab.

which is more tightly confined in the lower index of the two twin waveguides (almost always the III–V waveguide, due to the relatively high index of silicon), will enlarge as the III–V mesa is expanded, as shown in Fig. 4C. The addition of the silicon slab underneath the edges of the III–V mesa, which are coincidentally necessary for mechanical support, can cause the antisymmetric mode to radiate, as indicated by the oscillating field in the cladding in Fig. 4C. In Fig. 4D, the symmetric mode remains tightly confined and shows no sign of radiation loss.

Properly optimized dimensions of the III–V mesa and air trench widths can produce radiation losses which are extremely high, as much as 1 dB/μm. The lasers presented later in this chapter use a 24-μm-wide III–V mesa, with 0.8–1.6 μm silicon waveguides and 3 μm air trenches. The increased III–V mesa width can increase the threshold current by enlarging the amount of gain material which must be electrically pumped to transparency, but the simple addition of a hydrogen implant on the sides of the mesa to insulate the entire structure, except for a 4-μm-wide current channel centered over the silicon waveguide, mitigates the impact of this effect. Transparency current densities of around 100 A/cm^2/well can be achieved with this technique.

2.1.2 Loss Optimization

With this architecture for the cross section, there are a number of remaining free dimensional parameters that must be optimized: the width, thickness, and index of refraction of the III–V waveguide layers, the number of quantum wells to be used, and the dimensions of the silicon waveguide. Many elements of the design of the III–V waveguides can be copied from InP-substrate lasers, such as the quantum well and barrier alloy compositions, and the bandgap of the SCH layers (which is related to the index of refraction, with higher energy bandgap quaternary alloys having correspondingly lower index of refraction). In this work, a 1.2-μm bandgap wavelength was used because it offers a good trade-off between electrical and optical confinement.

The optical overlap with the III–V layers can now be controlled by the silicon waveguide dimensions (width and thickness) and the layer thickness of the III–V SCH. While the thicknesses of the silicon and III–V waveguide layers are fixed during the manufacturing of the SOI and InP wafers, the width of the silicon waveguide can be controlled by lithography, allowing different widths of silicon waveguides to be integrated in a single device. It will be shown in later sections that this is very useful for integrating high gain and high-power amplifiers, and for reducing the transition loss when transforming the mode from the amplifier waveguide into a passive silicon waveguide. Therefore, the silicon thickness and InP thickness and index should be chosen to produce a high-confinement factor, low propagation loss, and the largest possible confinement factor difference, $\Delta\Gamma_{xy}$, relative to the silicon waveguide width.

In the devices demonstrated in the following sections, a 500-nm silicon thickness was chosen early on as it represented a reasonable match to the effective index of common III–V waveguides. This, combined with the fixed bandgap in the SCH, reduces the optimization problem to a more manageable number of parameters: the silicon waveguide width and the III–V SCH thickness. It is likely that a different optimum design exists for different silicon thicknesses; for example, a thinner silicon layer might have optimal performance with thinner (or lower index) III–V SCH, and some earlier publications based on this technology used 700 nm silicon and higher index III–V layers (1.3 μm bandgap wavelength) (Fang et al., 2006b).

Optimization of the internal loss and confinement factor can be conducted with numerical simulation of the optical mode as long as the absorption coefficients and refractive indices for each material in the waveguide are known. Contributors to the propagation loss in this type of device are the absorption in the doped III–V cladding layers, side-wall roughness-induced scattering loss, absorption in the InGaAs p-contact (which has a bandgap wavelength of 1.65 μm), and radiation leakage loss, as discussed in Fig. 4. The scattering loss is measured in passive silicon waveguides to be below $0.5\,\text{cm}^{-1}$ for waveguides with dimensions in the range being discussed here, and so is neglected in this analysis.

Using calculated values of loss based on these simulations, and extrapolating the gain characteristics from small-signal measurements on Fabry–Pérot (FP) lasers, the net modal gain can be computed for a variety of waveguide dimensions (see Davenport et al., 2016 for the calculation of these parameters). Fig. 5 shows the results of these calculations. Maximum net modal gain increases with increasing SCH thickness, up to 25 dB/mm for 150-nm-thick SCH layers.

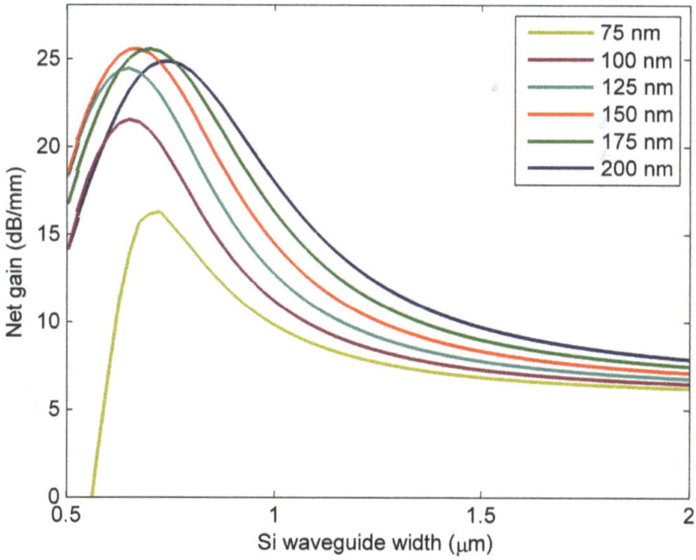

Fig. 5 Calculated net modal gain vs waveguide width for various III–V SCH thicknesses.

This is caused by increasing quantum well confinement factor, Γ_{xy}, which increases as the waveguide thickness decreases and SCH layer thicknesses increase as these cause the mode to be pulled up into the III–V mesa. For the thickest SCH layers, the mode begins to expand vertically out of the quantum well, reducing the maximum gain. Very narrow silicon waveguides likewise reduce the amount of lateral confinement, causing the mode to expand out of the pumped central part of the mesa into the unpumped hydrogen implanted sides of the mesa. For wider silicon waveguides, the mode begins to behave less like a twin waveguide and more like an evanescent waveguide, where the confinement factor depends solely on the III–V dimensions. Note that similar behavior would be observed for a fixed SCH thickness and varying SCH index.

Fig. 6A shows the impact of the SCH thickness on internal loss. Note that the legend refers to the thickness of the individual top and bottom SCH layer, so the overall thickness of the SCH is double what is displayed in the legend. Different SCH thicknesses result in optimal internal loss at different silicon waveguide width values, with thinner SCH layers being more optimal for wide waveguides and thicker SCH layers being more optimal for narrow waveguides. Overall, loss for thinner SCH layers is high when the silicon waveguide is narrow, where high-confinement factor would be expected. The internal loss of the thicker SCH layers is less variable. Note that for Fig. 6A the *y*-axis is a logarithmic scale; the loss for thin SCH layers and narrow silicon waveguides can be disastrously high because the leakage into the silicon slab is increased (Davenport et al., 2016).

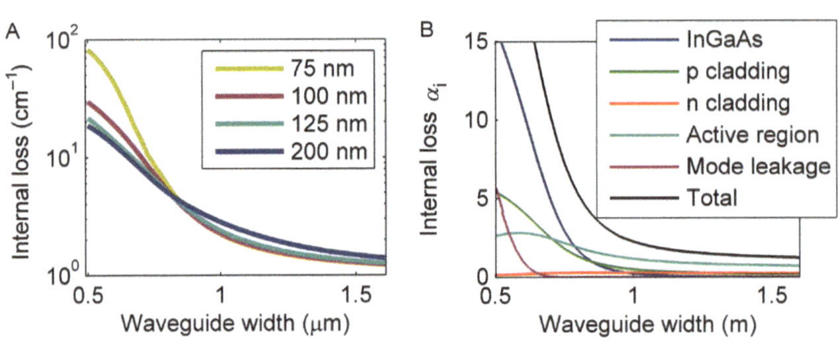

Fig. 6 Internal loss for various combinations of waveguide dimensions. (A) The variation of the internal loss vs waveguide width for differing III–V SCH layer thicknesses, and (B) The contribution of each individual factor for 125 nm thick SCH layers on the *top* and *bottom* of the quantum wells.

Fig. 6B shows the contribution from various loss factors for the 125 nm SCH thickness. Increasing confinement factor in the III–V layers with decreasing silicon width generally deteriorates the propagation loss. This is shown to be a result of absorption in the InGaAs as the vertical confinement is insufficient in the absence of the high-index InP lower cladding that is typically present in InP-substrate lasers. As the silicon waveguide narrows, the effective index of that layer decreases, creating an asymmetric mode that has higher overlap with the InGaAs layer. It is likely that a taller InP mesa and thinner InGaAs layer would reduce this source of loss to a level more in line with InP-substrate lasers.

Additional sources of loss are the absorption in the p-doped InP cladding. This design already uses a doping setback to help reduce this kind of loss; see Table 1. The n-InP cladding does not represent a major source of loss due to the fact that n-type InP is 10 × less absorptive than p-type InP, and the n-type InP layer is only 150 nm thick. For very narrow silicon lasers, the modal leakage due to radiation begins to increase rapidly. In addition, the carrier population in the quantum well produces some loss as well. This is the "active region" loss in Fig. 6B. It is expected that this value could be reduced by the introduction of larger compressive strain.

A similar analysis can be performed on the SCH thickness. Fig. 7 shows the simulated internal loss contributors for various SCH thicknesses at a silicon waveguide width of 850 nm. Thinner SCH layers display extremely high loss due to coupling with the InGaAs as they do not restrain the mode well in the vertical dimension. Thicker SCH layers, on the other hand, produce more radiation leakage loss, as the mode can more easily be coupled across the air trench into the silicon slab. There is a clear optimum value around 125 nm in Fig. 7. While the 150 and 175 nm SCH thicknesses had the highest modal gain, internal loss also affects laser slope efficiency, and so 125 nm was the value that was ultimately chosen.

2.1.3 Confinement Factor Optimization

The quantum well confinement factor is also a dominant factor in the performance of laser and amplifier devices, since it strongly controls the amount of gain per unit length that can be produced by the amplifier waveguide. Fig. 8 shows the impact of the lateral dimensions on the confinement factor. All designs have low-confinement factor for wide silicon waveguides, since the effective index of the higher index silicon material dominates and most of the mode is confined in that layer. For narrower silicon waveguide widths, the confinement factor increases up to a point where the mode begins

Table 1 The Design of the III–V Epitaxial Layers Used for These Devices, Along With Derived Refractive Index and Absorption Values

Epitaxial Layer Design

Layer	Material	Thickness (nm)	Doping (cm^{-3})	Bandgap Wavelength (μm)	Strain[a]	Refractive Index	α (cm^{-1})
1	$In_{0.53}Ga_{0.47}As$	200	p 10^{19}	1.65	0	3.7	10,000
2	InP	1000	p 1.5×10^{18}	0.92	0	3.17	33
3	InP	250	p 8×10^{17}	0.92	0	3.17	17.6
4	InP	250	p 5×10^{17}	0.92	0	3.17	5.5
5	$In_{0.53}Al_{0.18}Ga_{0.29}As$	125	p 1×10^{17}	1.20	0	3.38	1.1
6	$3\times$ $In_{0.68}Al_{0.06}Ga_{0.26}As$ $4\times$ $In_{0.44}Al_{0.09}Ga_{0.47}As$	6 9	u.i.d. u.i.d.	1.75 1.35	+1% −0.6%	3.41	83[b] —[c]
7	$In_{0.53}Al_{0.18}Ga_{0.287}As$	125	n 1×10^{17}	1.20	0	3.38	0.1
8	InP	110	n 1×10^{18}	0.92	0	3.17	1.0
9	$2\times$ $In_{0.85}Ga_{0.15}As_{0.33}P_{0.67}$ $2\times$ InP	7.5	n 1×10^{18} n 1×10^{18}	1.11 0.92	0 0	3.28 3.17	1.0
10	InP	10	n 1×10^{18}	0.92	0	10	1.0
11	Bonding SiO_2	7	N/A	N/A	0	1.45	0
12	Si	500	u.i.d.	1.10	0	3.47	0
13	Buried SiO_2	1000	N/A	N/A	0	1.45	0

[a]Positive means compressive strain, while negative values indicate tensile strain.
[b]This value varies with the carrier density. This is for high injection current (280 mA), as in SOA operation. For lasers, this value will be lower due to carrier clamping.
[c]Loss from the barrier is contained in the loss value used for the well.

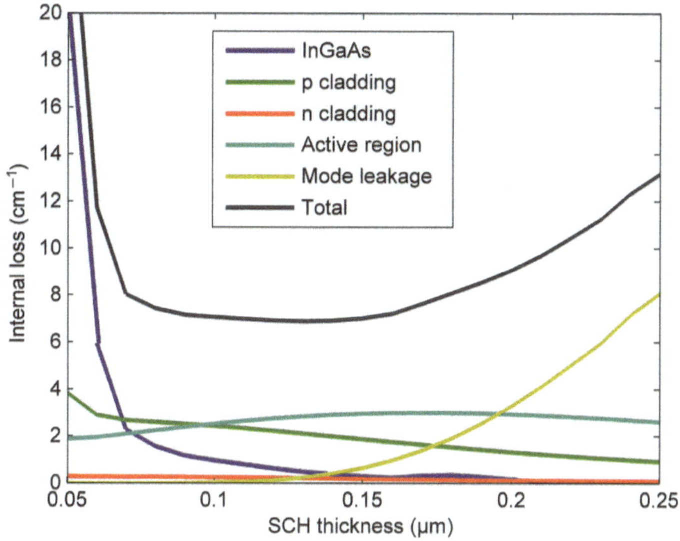

Fig. 7 Loss contributors for various SCH thicknesses for a fixed silicon waveguide width of 850 nm.

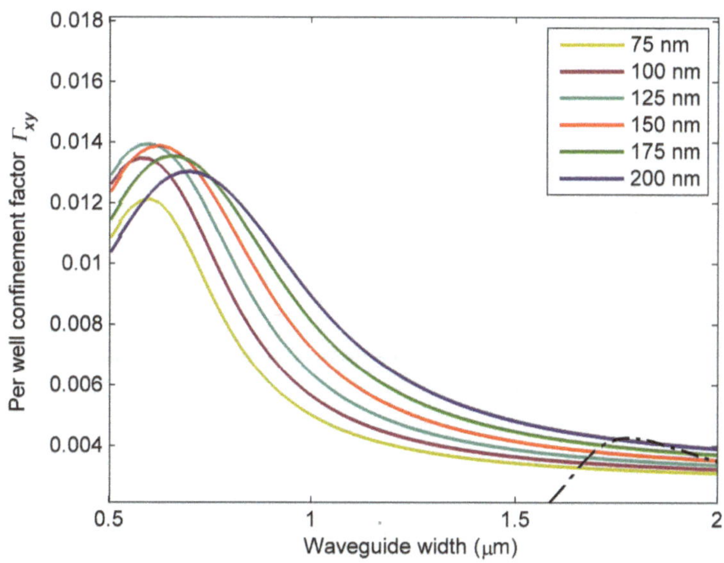

Fig. 8 Quantum well current channel confinement factor for the fundamental mode of the amplifier waveguide vs silicon waveguide width for various SCH thicknesses. The *dotted line* is the TE_{10} higher order mode for the 125 nm SCH.

to expand out of the pumped current channel region of the implanted mesa. As mentioned earlier, low-confinement factor can also be desirable, which gives an advantage to thinner SCH layers in this case. Note that the emergence of the first true higher order mode, the TE_{10} mode, limits the minimum attainable confinement factor, since this mode will reach threshold before the TE_{00} mode in cases where it has higher confinement factor.

The finalized epitaxial layer design is shown in Table 1. The alloy compositions, layer thicknesses, doping, bandgap, strain (where applicable), and the calculated refractive indices and absorption values are all listed in Table 1. More information and references for the computation of these values can be found in Davenport et al. (2016). All of the experimental data shown in subsequent sections are produced using this design or a close variant of it.

2.1.4 SOA Results

Variation of the quantum well confinement factor is normally difficult to accomplish in InP-substrate amplifiers. It is typically achieved with either variation of an epitaxial layer thickness or regrowth. In the case of heterogeneous silicon devices, as shown in Fig. 8, a range of confinement factors can be integrated on a single chip by simply varying a lateral dimension during lithography. The gain of an amplifier increases with increasing confinement factor, but the maximum output power decreases with increasing output power (Davenport et al., 2016).

A set of semiconductor optical amplifiers were fabricated using the waveguides design in Sections 2.1.2 and 2.1.3. The devices were 2 mm in length and were all fabricated on the same chip. They contain heterogeneous transitions (see Section 2.2) and output from passive silicon waveguides, which were antireflection coated to reduce reflection and prevent lasing. As expected, the narrowest silicon waveguide provides the most gain for low input power, 25.5 dB but saturates more rapidly (Fig. 9A). The lowest confinement factor waveguide, 1.6 μm, provides the highest saturation input power but the lowest gain (Fig. 9B). Ultimately, the 1.4 μm waveguide produced the highest output power, up to 16 dBm (Davenport et al., 2016).

In summary, the design of a heterogeneous silicon/III–V optical waveguide is a multivariable problem. The large number of degrees of freedom present difficulty but also the opportunity to finely optimize the performance of the device.

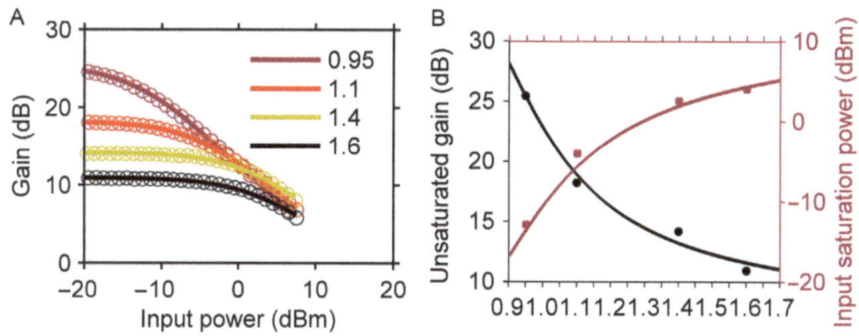

Fig. 9 The saturation characteristics of the heterogeneous silicon/III–V amplifier waveguide. The gain vs input power for various silicon waveguide widths (and therefore various confinement factors, with wider waveguides having lower confinement factor) is shown in (A). The *open circles* are measurement data, the *solid lines* are curve fits, and the *squares* represent the 3-dB saturation point extracted from the fitting. The fit parameters are plotted vs waveguide width in (B).

2.2 Active–Passive Integration

The heterogeneous amplifier waveguide described in the previous section allows for photonic circuit functions requiring direct bandgap materials to be integrated on a silicon photonic circuit. However, passive silicon waveguides are generally superior to InP waveguides in most respects, particularly bend radius and propagation loss. Unfortunately, the heterogeneous amplifier waveguide design has a considerably larger mode shape and a two-lobe mode profile that is quite unlike a typical single-mode silicon wire waveguide. Integration of the amplifier waveguide with a passive waveguide will require transformation of the mode to avoid creating high insertion loss and unintentional reflection.

High-performance laser devices will require high performance from the transition. For insertion loss, Fig. 6B shows that a 0.8-um-wide waveguide with the 125 nm SCH thickness has 5 cm^{-1} of internal loss, which translates to 2.2 dB of loss for a common 1 mm amplifier section. A transition loss of greater than 1 dB will then dominate the overall loss of the cavity, since the silicon waveguide loss is likely to be fairly negligible. Therefore, either longer amplifier sections should be used, or the loss of the transition should be <1 dB to ensure that it does not limit the performance of the device. A loss below 0.25 dB is unlikely to have a large impact on the efficiency of the device. Reflection should be below −40 dB to prevent the formation of parasitic FP resonances. Parasitic multimode emission, from too-abrupt transformation of the mode, must also be limited, as this can cause multipath interference if it reflected back into the transition by, e.g., a laser mirror. −20 dB is sufficient for single-wavelength lasers, but for broadband devices like tunable lasers or mode-locked lasers, −30 dB or lower is desirable. As with the insertion loss, these values represent an acceptable maximum, but will still often limit performance.

The amplifier waveguide is a twin waveguide in which, in the case of proper design, only the symmetric lowest-order mode is excited by the quantum wells. Transferring the mode from an active to a passive waveguide has been demonstrated on laser devices with a variety of methods, including a lateral directional coupler (Liang et al., 2011a), vertical directional coupler (Van Campenhout et al., 2007), grating-assisted codirectional coupler (De Koninck et al., 2013), or an adiabatic coupler (Ben Bakir et al., 2010).

Directional couplers are sensitive to wavelength and are extremely sensitive to dimensional variations introduced by fabrication error or design inaccuracy. In addition, for 100% coupling, the waveguides must be beta matched, which is difficult in the case of two highly dissimilar waveguides, such as the wide III–V slab and narrow Si ridge used in heterogeneous amplifier waveguides.

Grating-assisted codirectional couplers can achieve high coupling coefficients in the presence of beta mismatch but have extremely narrow bandwidth introduced by the grating structure. The grating also produces a parasitic reflection, which can be very low in theory but in practice is normally much higher than the −40 dB required by broadband laser devices.

The adiabatic coupler is normally extremely broadband (100 s of nanometers of wavelength bandwidth) and has low intrinsic parasitic resistance. "Adiabatic" in this case means that there is no coupling between modes, in contrast to the directional coupler, which launches the mode from the input waveguide into a superposition of modes in the coupled waveguide structure. Instead, the mode is gradually transformed into one of the modes (the symmetric mode in this case) of the coupled waveguide system.

2.2.1 Adiabatic Transition Concept

In the case of the heterogeneous transition from a device like the one described in the previous section into a passive silicon waveguide, the situation is not as complicated as lateral adiabatic couplers. The device is a 1×1 coupler, so it is more like a spot-size converter. However, the formalism of the adiabatic coupler is still useful for the design of the device.

The operation principle of the device is shown in Fig. 10. The first (from left to right) section of the device is the amplifier section, which

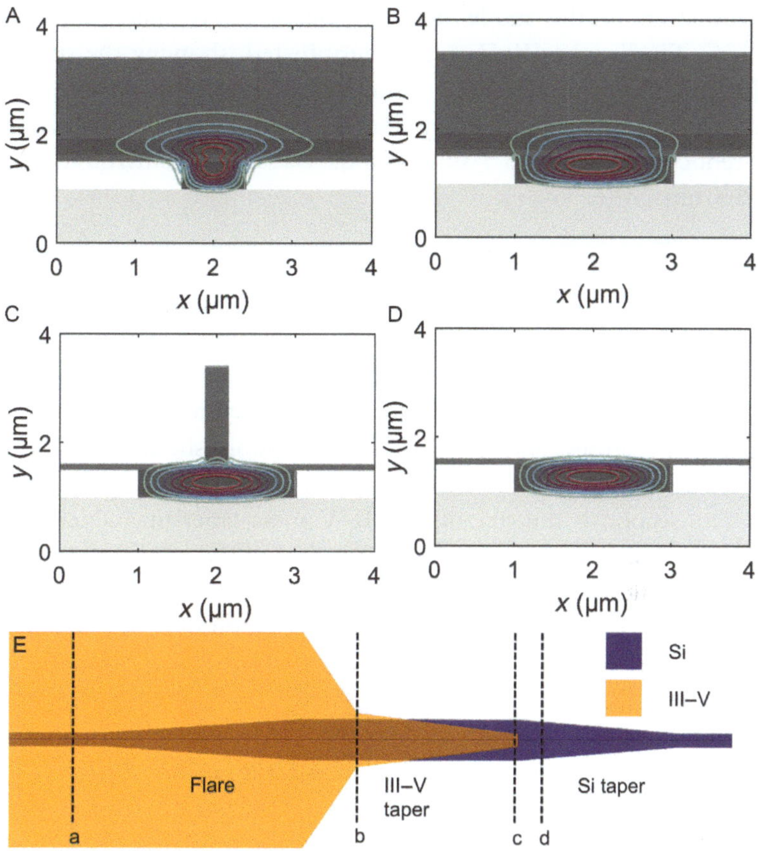

Fig. 10 Schematic demonstration of the adiabatic mode converter used to couple the heterogeneous amplifier waveguide into a passive silicon waveguide. (A)–(D) The transverse electric field profiles (contours) superimposed over the index profile (*gray scale*) of the waveguide cross sections of the slices indicated in (E).

may have a narrow silicon waveguide (Fig. 10A) for high-confinement factor in the quantum well. The silicon waveguide is then increased in width to reduce overall confinement in the III–V layers in the "flare." Then, the III–V mesa is rapidly reduced in width from the normal 26 μm width to 4 μm in the "neck" (this can be done over a short distance as the mode is not perturbed significantly by this change). The mode condition at the end of the neck is shown in Fig. 10B. The confinement factor in the III–V has been reduced significantly, often by a factor of 2 or 3. The loss through the flare is normally minimal, since it can be quite long, 100 μm normally. At this point, the III–V mesa is tapered down to a point, completing the transformation of the mode into the silicon waveguide, as shown in Fig. 10C. Finally, the III–V ridge is terminated, showing the single-lobe fundamental TE_{00} mode that is required for matching with the passive silicon circuits following the amplifier section. The silicon waveguide is then reduced in width to a single-mode waveguide dimension, typically 600–800 nm.

2.2.2 Loss

Loss in the transition is highly undesirable. It reduces the output power and electrooptical efficiency of laser devices and elevates the threshold carrier density, which in turn impairs the high-temperature performance of lasers and amplifiers. The loss of the device comes down to a number of conditions. Intrinsic losses, like scattering or absorption, should be minimal, as the overall length of the device will cause them to be a significant contributor. The second is the size of the III–V mesa taper tip, which must be minimized to prevent significant perturbation of the mode at termination; 700 nm is sufficient for low loss (but not low reflection, as described in Section 2.2.3. The third is the length of the transitions: abrupt transitions create loss and emission into higher order modes, and for this type of coupler to be truly adiabatic, the length of the lateral tapers needs to be sufficient. The loss asymptotically approaches 0 as the length approaches infinity. However, longer structures are subject to increased loss from other sources, such as absorption and scattering loss, so the length of real devices often has to be limited to balance loss from the mode transformation and loss from other sources.

The flare and the passive silicon taper can be elongated, since they have relatively low propagation loss (in fact, the flare has net gain). However, the III–V mesa taper is a different story. It is deeply etched (1.8 μm in total, see Table 1), which introduces some sidewall roughness, and the narrow III–V structure can be difficult to drive sufficient current for transparency. Indeed, the dry etch technology used in this work (methane–hydrogen–argon reactive ion etching) to produce the deep III–V mesa contaminates the thin taper structure and reduces its conductivity even more than would be expected by the small lateral dimensions (Kreissl et al., 2000). As a consequence, the III–V taper section must be as short as possible. A wide silicon flare (1.8–2.0 μm), thick silicon layer (500 nm is a reasonable minimum for 1550 nm wavelength operation, this scales with wavelength), and properly designed III–V layers to achieve low-confinement factor are all critical for reducing the length of the III–V taper. Ultimately, good performance from III–V tapers in the 20–30 μm range was attained. These transitions exhibited between 0.25 and 1 dB of loss depending on the wavelength, due to absorption (Davenport et al., 2016). Simulation suggests that the loss due to nonadiabaticity is 0.2 dB, meaning that 0.75 dB at the point of highest absorption (the blue side of the emission spectrum) is due to absorption and scattering. Electron micrography of the device does not suggest that scattering is a large contributor, as the sidewalls appear quite smooth. This value could likely be reduced by using a hydrogen-free etch process, such as chlorine-based reactive ion etching, or a hydrochloric acid-based wet etch, as used by Keyvaninia et al. (2013a).

2.2.3 Reflection

Low reflection is also critical for laser devices. Unintended reflections create parasitic cavities that disrupt the operation of integrated filters in single-wavelength lasers or reduce the operating bandwidth of broadband lasers by creating interference. The tall III–V mesa, which normally has refractive index greater than 3, is clad with dielectric such as silicon dioxide or silicon nitride, with refractive index less than 2. This creates the potential for a large Fresnel reflection at the termination of the taper tip. This can be reduced by simply reducing the size of the III–V mesa. In theory, a three-dimensional taper would be ideal, but because the device requires electrical contact to prevent absorption, this is impractical. Instead, the mesa must be tapered to very narrow lateral dimensions.

FDTD simulation indicates that −40 dB reflection can be attained with taper tip size below 500 nm, and −50 dB by tip sizes below 300 nm. Both of these values are challenging to achieve in very tall features, as the tip will have an aspect ratio of up to 6:1 in the case of the 300 nm taper. In reality, only the active region layers (the SCH and quantum wells) taper needs to be this small, since the majority of the mode is contained in these layers. It was found that the optimal solution is to make a wider InP mesa, which constitutes most of the height of the III–V ridge, and then use the wet etch for the quantum well and SCH layers (Section 1.1) to undercut the active region, producing a dead-sharp taper tip for those layers, which is guaranteed to be electrically contacted by the larger overhanging p-type InP cladding.

Measurement of the taper reflection can be conducted by examining the output spectrum of below-threshold FP lasers using the method of Barbarin et al. (2005). Measurements from devices with 500-nm-wide p-InP cladding and ~200 nm wide active region tapers are shown in

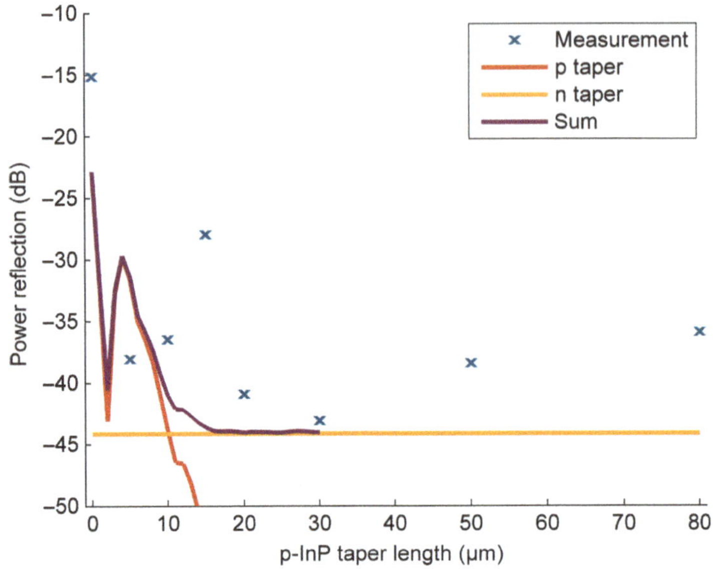

Fig. 11 Measurement and simulation of reflection from heterogeneous transitions. *X* symbols indicate measurement data. The *red line* is the reflection from the III–V mesa, and the *yellow line* is the reflection from the thin n-layer transition (to be discussed in the next section). The *purple line* is the sum of those reflections.

Fig. 11, vs the length of the taper. The measurement data are shown as symbols, while the lines are FDTD simulations. The simulations qualitatively represent the data: the large increase in reflection for short tapers, where the transition is nonadiabatic so that the mode is still confined in the III–V mesa, and the spike around 15 μm in length, which represents a corner reflector effect, are both present in the simulation but not precisely where they appear in the measurement data. The floor in the reflection is caused by the thin n-type InP contact layer transition, which will be discussed in the next section. The increased reflection for longer tapers is likely simply due to increased noise floor in the measurement as the longer tapers have significantly higher loss due to absorption. It is impractical to use FDTD for simulations of devices that size and other methods (eigenmode expansion, beam propagation techniques) do not represent the parasitic reflection accurately.

2.2.4 N-Layer Transition

The thin n-layer transition, while seemingly a lesser challenge due to its factor of 10 lower overall thickness compared to the active region and p-type mesa, still presents some difficulty. Because of the height of the nearby III–V mesa, precision lithography is much more difficult at this stage of the fabrication (see Fig. 2) because thicker photoresist must be used to cover the tall III–V mesa. In addition, etching the n-layer exposes the air trench of the silicon waveguide, so using a wet etch to sharpen the tip of the taper is inadvisable since the wet etchant can penetrate the trench and attack the InP from the underside.

In addition, the nature of the n-layer taper makes it more difficult to fully remove the optical field overlap from this layer before introducing a termination. The III–V mesa is a twin waveguide system, and the upper lobe of the symmetric mode can be shrunk until it is negligible; however, the n-layer always has lower index of refraction than the silicon, so it is essentially just a cladding layer that the optical field will decay evanescently into. Because of this, low reflection can only be achieved if the tip size is impractically small, below 200 nm.

Also, because the mode profile is essentially similar in the waveguide with the n-layer compared to the waveguide without the n-layer, coupling to higher order modes in the silicon waveguide, which is now enlarged to be highly multimode in the flare, is more likely. The magnitude of this effect is shown in Fig. 12. For perfect alignment, the parasitic higher order

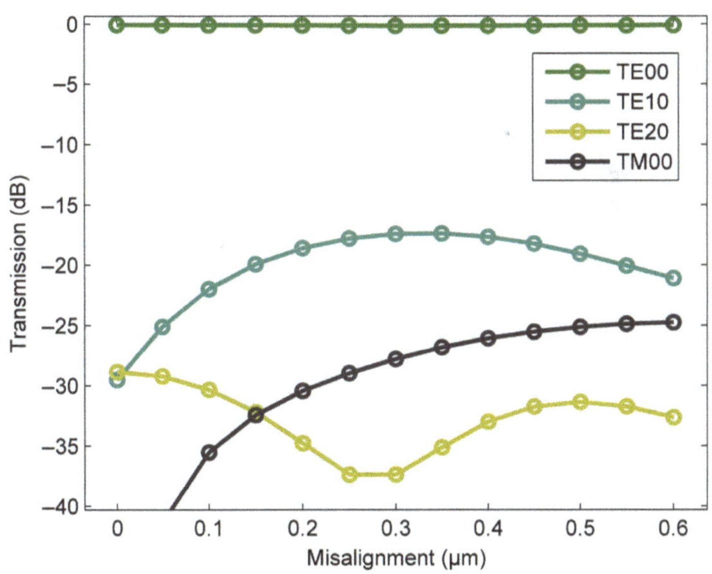

Fig. 12 Simulated transmission into the most significant higher order modes by an n-type InP when it is misaligned due to lithography error.

mode transmission is acceptable, at −30 dB. However, for even small misalignment, this transmission increases to worse than −20 dB, which is unsuitable for most laser applications.

Using a more sophisticated lithography technique with tighter tolerances and smaller feature sizes (this work was mostly conducted using an i-line projection lithography tool for the III–V fabrication, with the exception of the III–V mesa). With that being generally unavailable in III–V fabrication facilities, an alternative is needed. Fig. 13 shows an angled junction, where the silicon waveguide is crossed at an angle by the n-InP layer. This eliminates the need for a small tip, as the tip of the triangle is far away from the waveguide, and it is impervious to misalignment. The angle reduces reflection, and with proper optimization, it can have transmission into higher order modes below −30 dB. Angles above 15 degrees exhibit below −50 dB parasitic reflection. This is superior to taper designs even in the ideal case unless the taper can be sharpened to below 150 nm.

FDTD simulation of the parasitic higher order mode transmission is shown in Fig. 14A. For 45 degrees, the transmission into the two

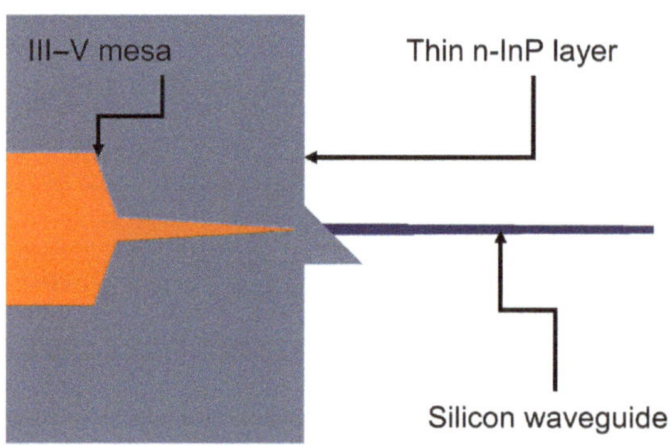

Fig. 13 Plan-view schematic of the angled n-layer transition.

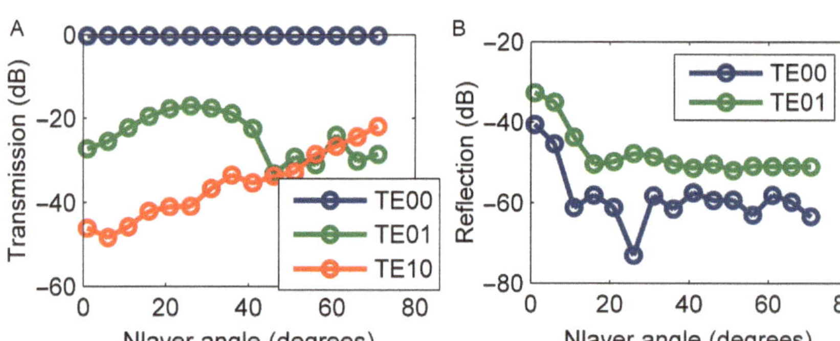

Fig. 14 Simulated higher-order mode transmission (A) and reflection (B) for the angled n-layer transition.

propagating modes with the highest excitation is right below −30 dB. Fig. 14B shows the parasitic reflection of the device vs angle. Any angle above 15 degrees has negligible reflection; however, there is a significant reflection into the TE_{01} mode which reaches an acceptable level at a higher angle. The results for high-performance integrated tunable and mode-locked lasers presented in Sections 3.3 and 3.4 use this angled n-type transition.

Low loss and low parasitics are essential for enabling high-performance integrated circuit lasers that combine active and passive waveguide devices. This design for a heterogeneous transition, based on a 1×1 adiabatic coupler concept, can produce loss below 1 dB, reflection below −40 dB, and higher order mode emission at −30 dB, and is suitable for use in high-performance integrated laser devices.

2.3 Cavity Design

With the heterogeneous amplifier and transition addressed in Sections 2.1 and 2.2, the heterogeneous silicon/III–V device is now prepared to leverage one of the strengths of the silicon photonics platform: superior passive waveguides. The laser reflector can be formed from the silicon waveguide, taking advantage of the high-confinement, low propagation loss, and high-precision fabrication available in this platform.

Laser mirrors come in two broad categories. Broadband mirrors are useful for lasers which emit multiple wavelengths simultaneously, such as comb lasers and mode-locked lasers, or for lasers where narrow spectral emission is less critical, like pump lasers. Single-wavelength mirrors, where the reflection is limited to a narrow wavelength bandwidth, are used in tunable and fixed single-wavelength lasers. Both of these types of mirrors can be produced on the silicon photonics platform.

2.3.1 On-Chip Broadband Mirrors

Broadband mirrors can be produced with a variety of techniques. The distributed Bragg grating reflector is one example, particularly on silicon, where deeply etched gratings can achieve extremely high kappa owing to the high-index contrast available in the system. These types of mirrors are common in mode-locked lasers based on heterogeneous silicon/III–V technology (Keyvaninia et al., 2013a). One drawback of this technology is that it requires extremely small feature sizes, below 130 nm for gratings centered around 1550 nm wavelength. An alternative is the Sagnac loop mirror, which was originally intended for use in fiber optics as a nonlinear

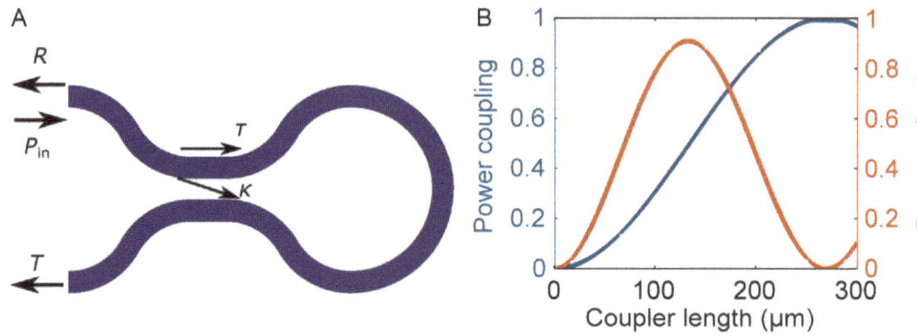

Fig. 15 Basic principle of a loop mirror. (A) A plan view of the device, with the input waveguide in the *upper left*, the directional coupler in the *center*, the loop on the *right side*, and the output waveguide on the *lower left*, and (B) The impact of the directional coupler length on the behavior of the mirror.

optical switch (Doran and Wood, 1988), and was later demonstrated on silicon for use in on-chip interferometers (Sun et al., 2013).

The loop mirror is essentially a 2×2 coupler, usually a directional coupler, with the two output waveguides connected with a loop of waveguide. This is shown schematically in Fig. 15A. The reflection can then be represented by the simple formula $R = 4\kappa\tau e^{-2\alpha L}$, where R is the power reflection, κ and τ are the power coupling and transmission coefficients of the coupler, α is the propagation loss in the waveguide, and L is the overall single-pass length of the device, including the length of the directional coupler and the loop. Fig. 15B shows the simulated directional coupler coupling coefficient vs length (blue curve) and the calculated power reflection, assuming a fairly large (300-μm long) and high loss (10 dB/cm) waveguide in the loop. Even with these pessimistic values, high reflection is still achievable. Using adiabatic bends, small bend radii, and small directional coupler gaps to reduce the overall length of the device could realistically reduce the overall length of the mirror to less than 100 μm.

The wavelength bandwidth of this type of mirror is limited by the wavelength dependence of the directional coupler, simulation of which is shown in Fig. 16. Short couplers with lower coupling coefficients have increasing reflection coefficient with increasing wavelength, as the mode expands out of the waveguide, increasing the evanescent coupling in the coupler gap. This makes the reflection simply increase with increasing wavelength, so there is not a bandwidth per se. For maximum reflection mirrors, where $\kappa = 0.5$, this increased coupling due to increasing wavelength actually reduces the reflection, giving very broad bandwidth. The worst-case scenario is for an $R = 0.5$ mirror, which only varies by 0.045 over a practical mode-locked laser bandwidth of 11 nm. Replacing the directional coupler with a 2×2 coupler with broader bandwidth, such as a curved coupler or an adiabatic coupler, would resolve the bandwidth limitation on partially reflecting mirrors for applications that demand wider wavelength range, like a widely tunable laser.

The continuous adjustability of the reflection allows for fine optimization of laser mirror reflectivity for high-power lasers. One example of a multiwavelength laser based on loop mirrors, with a 50% reflection loop mirror on the back of the laser and a 15% reflection loop mirror on the output side, is shown in Fig. 17. The light–current characteristic shows a

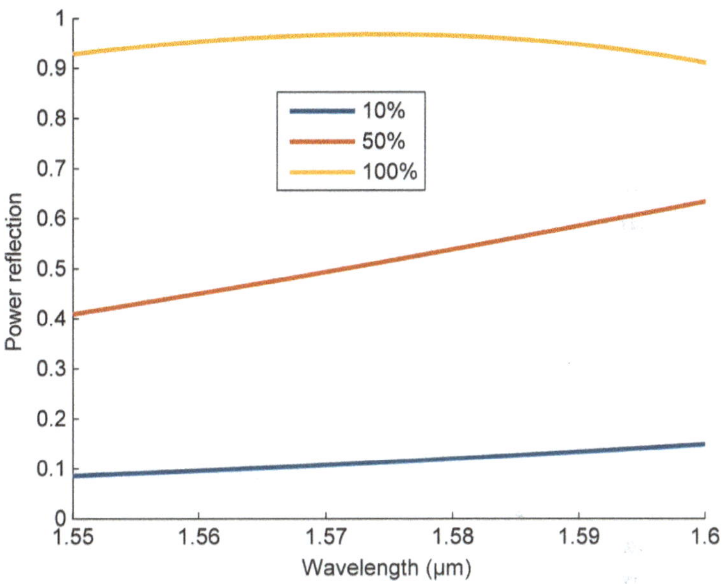

Fig. 16 Wavelength dependence for three loop mirrors with various power reflection values, indicated by the legend.

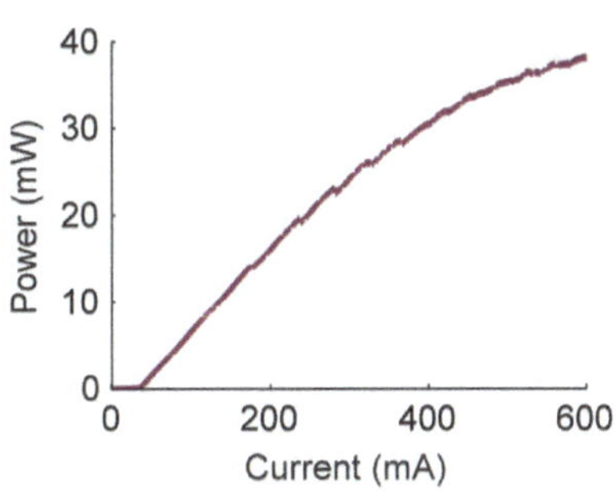

Fig. 17 Power vs current curve for a loop mirror-based laser, detected from a single side with an integrating sphere.

low threshold of around 30 mA and very high output power of 38 mW, which is exceptional for a heterogeneous silicon laser. The device has two passive loop mirrors and a central heterogeneous amplifier with two heterogeneous transitions inside of the cavity. Light was collected from an angled and antireflection-coated facet with an integrating sphere, so this represents the on-chip power.

2.3.2 Single-Wavelength Filters

As the laser cavity length is increased, the longitudinal modes get closer to each other in frequency domain. In a typical semiconductor laser (e.g., 1 mm long laser, group refractive index $n_g=3$) that spacing is in the order of 50 GHz. For single-frequency operation, it is necessary to include a filter capable of filtering a single longitudinal mode across the bandwidth of the gain medium which is in the order of 10 THz (80 nm at 1550 nm). The single-frequency purity is commonly described with parameter called mode suppression ratio (MSR) which relates the power in the primary laser mode to that in the next strongest mode from one end of the laser. In a good laser the MSR is greater than 40 dB and is achieved with the combination of passive (cold cavity) and active filtering. For a stable laser, commonly the modal gain difference should be >2 dB and this difference is then amplified by multiple passage of light through the laser cavity to provide >40 dB MSR.

For single-frequency lasers, one can utilize Bragg gratings in the gain medium to realize distributed feedback (DFB) lasers. Combining filter elements into the gain section is an effective solution to shrink the size of the laser, allowing for low threshold current and high-speed direct modulation (Nakamura et al., 1973; Zhang et al., 2014). Since the optical mode in heterogeneous waveguide hybridizes in both III–V and Si (as shown in Section 2.1), the Bragg gratings can be defined solely on silicon waveguides in the silicon process prior to bonding. This is in fact a valuable advantage of the heterogeneous silicon platform over the conventional III–V monolithic platform where the formation of Bragg gratings often involves epitaxial regrowth or other complex processes. As lithography and etching on silicon can be done at large scale with perfected processes inherited from the mature silicon electronics industry, the fabrication of the Bragg gratings can be obtained with high precision; therefore, heterogeneous Si DFB lasers offer high-precision control of the lasing wavelength, which is of critical importance in many applications.

The heterogeneous silicon photonic platform also enables a new degree of freedom to control the coherence (or equivalently, the spectra linewidth) in the DFB lasers. By lowering the mode confinement in III–V and having the optical mode localize more in silicon passive waveguide, one effectively reduces the spontaneous emission (which happens in the III–V active region) coupling rate to the lasing mode. In addition, the internal loss of the mode can be reduced and hence gives room for designers to achieve a cavity with quality (Q) factor that is orders of magnitude higher than achievable in conventional III–V monolithic laser. Heterogeneous Si DFB lasers with record low sub-kHz spectral linewidth have been successfully demonstrated with this approach (Santis et al., 2015). However, as a higher Q cavity also means lower mirror loss, there is an inherent trade-off between the coherence and the output power of the DFB laser operating in this scheme. As shown next, this trade-off can be eliminated with the passive resonator approach.

Wide tunability is commonly achieved by utilizing the Vernier effect. The effect has been utilized both with sampled Bragg grating reflectors and ring resonators. Ring resonators, provided that the utilized waveguide platform offers sufficiently low propagation losses, have an advantage as the effective cavity length at ring resonance is significantly enhanced, directly reducing linewidth (Liu et al., 2001, 2002). A single ring resonator in add–drop configuration placed inside the laser FP cavity could be used for selecting an individual longitudinal mode. A typical transmission curve of a single ring resonator is shown in Fig. 18A and is periodic; where the period or free-spectral range (FSR) corresponds to the inverse round-trip

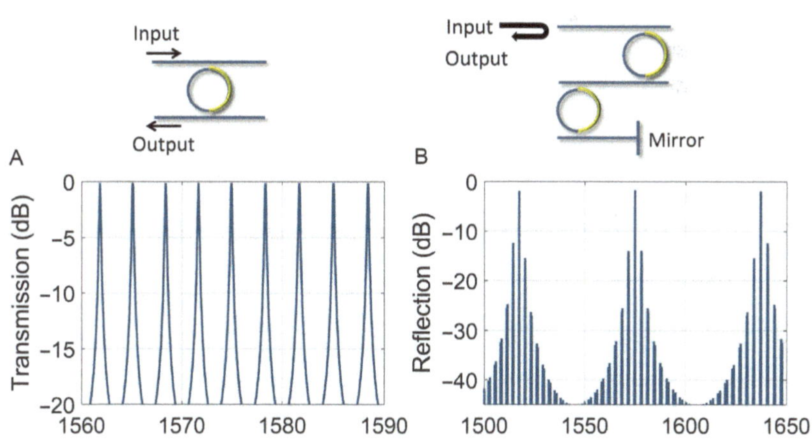

Fig. 18 (A) Add–drop response of single ring resonator ($L=200\,\mu m$). (B) Add–drop response of two rings with slightly different radius ($L_1=200\,\mu m$, $L_2=211\,\mu m$) utilizing Vernier effect and providing much wider single-frequency response.

time inside the ring. To guarantee single longitudinal mode operation, the FSR of the ring should be larger than the gain bandwidth of the laser. In our exemplary case of 10 THz gain bandwidth, the ring circumferences should be as small as 10 μm in Si which might be impractical for at least two reasons. First, the light has to be coupled to the ring at two points, and such short couplers are challenging. Second, we want to provide wideband tuning; ideally across the full gain bandwidth of the semiconductor material. Tuning in silicon photonics is commonly performed via two mechanisms: thermal tuning due to primarily refractive index change of material due to temperature change (dn/dT) and carrier tuning based on carrier plasma dispersion effect. The latter is related to the density of free carriers in a semiconductor, which changes both the real and imaginary parts of the refractive index. The carriers can be injected or depleted. Injection is more efficient in terms of tuning efficiency but is slower and consumes more power compared to depletion. In both cases, the change of carrier concentration results in loss modulation, which is undesirable. Thermal tuning, on the other hand, has minimal effect on loss and is preferred for our applications where low noise is crucial. The disadvantages are relatively high-power consumption and low speed of tuning, and there is a direct trade-off between the two, where if one optimizes for lower power, it leads to lower tuning speed or vice versa. The dn/dT of silicon is around $1.87 \times 10^{-4}\,\mathrm{K}^{-1}$ at 1550 nm and room temperature (Frey et al., 2006) which translates to a tuning efficiency of ~0.08 nm/°C at 1550 nm. Covering the whole gain bandwidth would therefore require as much as 1000°C of temperature change, if the silicon dn/dT is constant, which it is not. In practice, up to 20 nm of tuning has been demonstrated with passive devices, but as laser performance is significantly influenced by temperature—minimizing the required temperature change is preferred.

Utilizing the Vernier effect with two or more rings with slightly different circumferences allows drastic reduction in required temperature change. In a typical widely tunable laser, two or more different rings are designed so that their FSR is between 200 and 400 GHz. Such rings can be tuned by full FSR with only 20–40°C of tuning, and due to slight difference in circumference, their resonances align only at particular wavelengths so to provide single-mode operation in much wider bandwidth than the FSR of each individual ring, as shown in Fig. 18B (Komljenovic et al., 2017).

The use of rings inside the laser cavity benefits the linewidth for two reasons. One is the reduced longitudinal mode confinement, which is maximized at ring resonance, and the second one is the reduction in linewidth from the negative feedback effect, where a decrease in wavelength increases reflectivity (increasing photon density in the cavity), and hence decreases carrier density, which in turn causes the wavelength to increase, due to the carrier plasma effect (Kazarinov and Henry, 1987). The combination of low propagation loss and high-quality (Q) factor rings allows substantial improvement in laser linewidth down to sub-10 kHz range for fully integrated semiconductor lasers, which is around three orders of magnitude improvement compared to initial III–V widely tunable lasers realized with sampled distributed Bragg gratings.

3. LASER DEMONSTRATIONS

Various laser structures have been demonstrated in the heterogeneous photonics platform with exceptional performance. Such lasers can readily be integrated with other components to form more complex PIC devices, as they are based on on-chip loop mirrors to form the laser cavity (Hulme et al., 2015, 2017; Tran et al., 2017; Zhang et al., 2016). In the remainder of this section, we give an overview of recent results.

3.1 FP Lasers

The FP laser, which consists of a simple broadband mirror at either output of an optical amplifier, is the simplest type of laser. In diode lasers they are commonly used as high output power pump lasers, since the cleaved semiconductor facet is typically very close to an ideal mirror with no excess loss, unlike on-chip mirrors. In addition, the FP laser is helpful for evaluating the properties of an amplifier waveguide, since it separates the performance of the active waveguide from the performance of its integrated mirror.

Heterogeneous silicon lasers are typically accompanied by a heterogeneous transition, which makes them useful for integration with silicon photonic circuits. However, as explained in Section 2.2, this component is very complicated and can confound attempts to analyze the performance of the laser, since the impairments introduced by the transition are typically challenging to separate from the performance of the amplifier section. Therefore, it is helpful to produce a classic pure FP laser when evaluating heterogeneous lasers.

Unfortunately, this task is complicated by the fact that silicon is much more difficult to cleave than the III–V semiconductors. The material is harder and tends to cleave into diagonals when the common (100) orientation silicon wafer is used. Furthermore, the bonded III–V material may not line up precisely with the crystal orientation of the silicon. Some report that it is possible to produce a heterogeneous facet with cleaving (Santis et al., 2014), but it often has poor yield due to the large force required to cleave the silicon.

Instead, mechanical polishing of the facet can be performed to produce a pure FP laser with heterogeneous facets. This is conducted with a flat polishing platen and a serially decreasing set of polish film grits on a laser bar that was cut with a dicing saw (as is typical for silicon substrates). Polishing is normally performed with equipment meant for polishing fragile materials such as transmission electron microscope cross sections or fiber optic ferrules. Polishing films using diamond grits are preferable, although very fine grit films with diamond are often unavailable and alumina or silica can suffice. A polishing film series starting with 6 μm grit and proceeding through 1, 0.5, 0.3, and 0.02 μm films will produce a facet surface with 5–10 nm RMS of surface roughness, which is sufficient for repeatable-performing FP lasers.

Preparing pure FP lasers allows for a number of useful techniques. Spectral gain measurements allow for analysis of the properties of the amplifier without being confounded by the device length; for example, very high gain amplifiers often have high transparency current, which can produce high threshold current lasers. While high threshold can indicate problems with the laser design, high threshold accompanied by very high gain simply means that the laser cavity length can be decreased to simultaneously improve threshold (by reducing transparency current), increase slope efficiency (by increasing mirror loss), and reducing device size. An example of an early heterogeneous silicon/III–V laser design, functionally equivalent to the first published result of an electrically pumped laser on silicon (Fang et al., 2006b) was characterized using Cassidy's method to extract the gain–current relationship (Cassidy, 1984). The resulting gain spectral gain is shown for various bias currents in Fig. 19A. These spectral gain plots contain a wealth of information about the performance of the laser. Multiple transverse mode lasing would manifest itself as modulation on the spectrum, since the higher order modes would not couple into the single-mode fiber used to collect the light for this measurement, manifesting as loss at wavelength positions where the two modes interfered destructively.

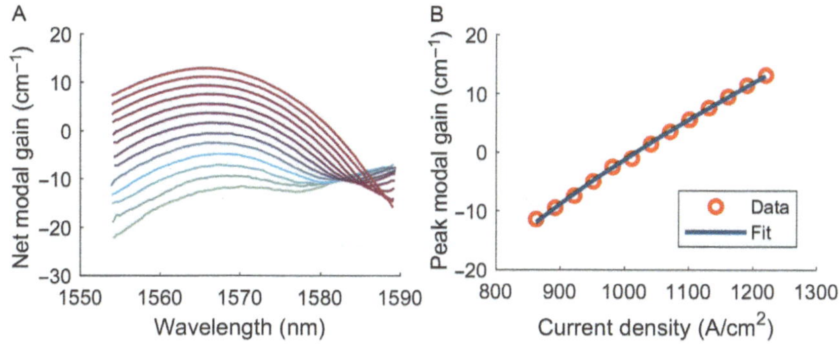

Fig. 19 Gain spectra (A) and fitted gain–current relationship at the gain peak (B).

The decreasing gain/increasing loss at very red wavelengths signifies that the buildup of carriers in the active region is creating increasing loss at wavelengths outside of the gain bandwidth, which indicates insufficient compressive strain in the quantum well (Fuchs et al., 1992).

The gain values at the lasing wavelength are shown in Fig. 19B as red circles. The empirical gain–current relationship, $g = \Gamma g_0 \ln(J/J_{tr})$, is fit to the measurement data as a blue line. This fitting revealed a Γg_0 value of $72\,cm^{-1}$ and J_{tr} of $890\,A/cm^2$, which signifies that this eight quantum well design epi simply had too much gain for the scale of devices that were being made using this design. Adjustment of the design based on this information, specifically reducing the number of quantum wells and increasing the compressive strain, was important for improving the performance of heterogeneous silicon/III–V lasers based on this technology.

3.2 Distributed Feedback Lasers

Since the first DFB laser on heterogeneous silicon photonics platform was demonstrated (Fang et al., 2008), various DFB lasers have been designed for different targeted performances (e.g., Santis et al., 2014; Srinivasan et al., 2011; Zhang et al., 2014). In this section, we present a simple design, compact (650 μm long) phase-shifted DFB laser that has more than 20 mW maximum single-sided output power and down to 400 kHz spectral Lorentzian linewidth.

The schematic cross section of the laser is shown in Fig. 20A. The silicon rib waveguide is 1.2 μm wide, 500 nm thick with 231 nm deep rib patterned with DUV lithography and dry etch in a DRIE etcher. With that geometry, the confinement factor in the QW region is estimated about 5%. More than 50% of the optical mode is confined in the silicon waveguide.

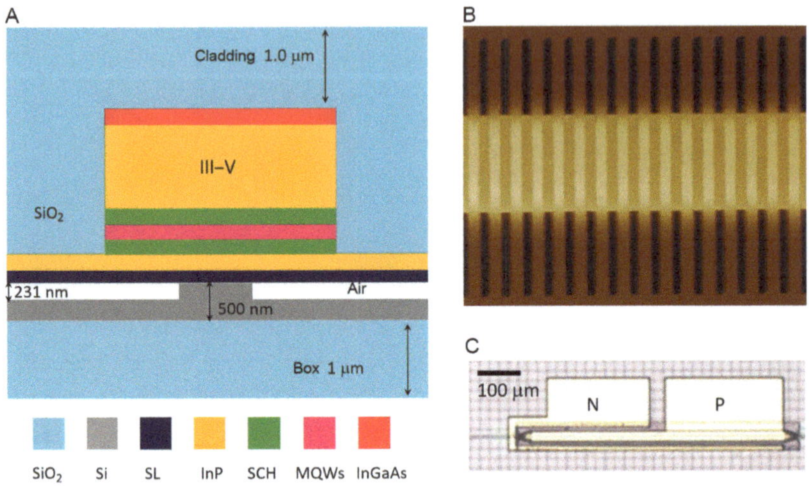

Fig. 20 (A) Schematic cross section of the DFB laser. (B) AFM image of the surface grating on silicon waveguide. (C) A microscopic photograph of a fabricated DFB laser on heterogeneous silicon photonics.

First-order uniform gratings with a quarter-wave phase shift at the center of the cavity were defined on the silicon waveguide surface using an electron beam lithography system and then etched with an ICP etcher. The grating etch depth was only 3 nm, and an AFM image is shown in Fig. 20B. The fabrication is completed with III–V bonding and subsequent heterogeneous processes shown in the previous sections. Heterogeneous transitions and tapers were formed on two ends of the laser, transferring the modes to single-mode passive silicon waveguides. A microscopic photograph of a completed laser structure is shown in Fig. 20C.

The LI curve in Fig. 21A shows 25 mA threshold current and maximum 22 mW single-sided output power. The lasing spectrum at 150 mA is shown in Fig. 21B. The laser spectral linewidth is measured with the delayed self-heterodyne method (Coldren et al., 2012). A measured spectrum is plotted with a Lorentzian fit in Fig. 21C of a 400-kHz spectral linewidth. Finally, the linewidth vs inversed output power was measured, showing expected linear relationship above threshold. The linewidth saturates and rebroadens at high injection current level possibly due to a combination of gain nonlinearity, thermal, and spatial hole-burning effects.

3.3 Widely Tunable Lasers

Early generations of heterogeneously integrated ring-based Vernier lasers featured linewidths typically in the range of 50–300 kHz (e.g., Komljenovic

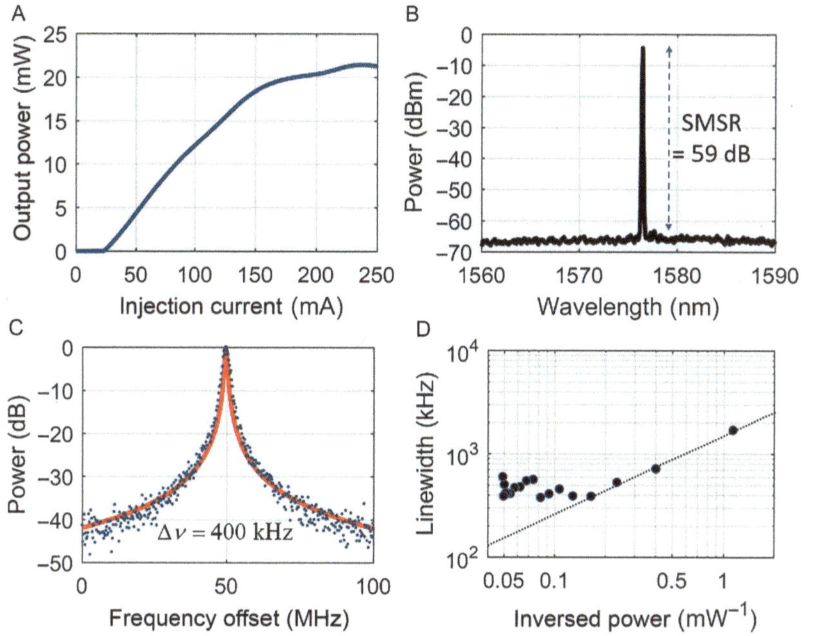

Fig. 21 (A) Single-sided output power vs injection current. (B) A lasing spectrum shows near 60 dB side mode suppression ratio. (C) Measured RF spectrum of the laser in a delayed self-heterodyne measurement with a *Lorentzian fitted curve.* (D) Laser spectral linewidth vs inversed output power.

et al., 2017; Srinivasan et al., 2015). More recently, with careful optimizations of silicon waveguides as well as the III–V gain layer designs, we have been able to further improve the linewidth by an order of magnitude. Here, we present a laser structure that demonstrates 55 nm tuning range, about 55 dB side MSR and record narrow spectral linewidths of 2–6 kHz across the C band.

The laser schematic structure is shown in Fig. 22A. The front-side mirror is formed by an integrated loop mirror with a 2 × 2 directional coupler. The back-side complex mirror is another loop mirror formed by a broadband 50% 1 × 2 multimode interferometer coupler. Within this loop, two add–drop ring resonators are bridged by an unbalanced Mach–Zehnder interferometer (UMZI) filter. The FSR of the UMZI is designed to be exactly half of the average FSR of the two rings. Using the heater on the longer arm, the UMZI spectrum can be fine-tuned to have its maximum aligned at the lasing wavelength—the coincident wavelength of the two Vernier ring resonances—while the minima are aligned at the neighboring

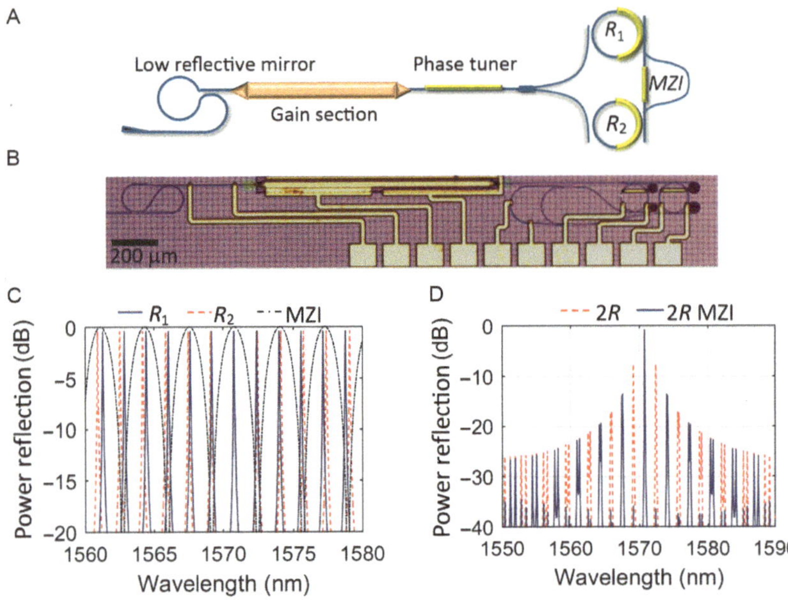

Fig. 22 (A) The schematic structure of the tunable laser. (B) Microscopic photograph of the fabricated laser. (C) Simplified theoretical individual reflection spectra of the two rings and UMZI. (D) The synthesized reflection spectra of the two-ring mirror with and without the UMZI, showing >7 dB enhancement in SMSR over the tuning range with the inclusion of the UMZI. The waveguide dispersion is ignored for simplicity.

resonant wavelengths. As can be seen in Fig. 22C and D, the addition of the UMZI significantly increases the passive SMSR over the tuning range by completely suppressing the first side modes.

A fabricated device occupies a small area of 1.1 mm^2 (Fig. 22B). The passive Si waveguides were all quasi-single-mode rib waveguides with the average propagation loss of 1 dB/cm. At room temperature, the laser showed a threshold current of about 33 mA and more than 16 mW optical output power on-chip at 150 mA current injection, as can be seen in the LI curve (Fig. 23A). The wavelength tunability with 100 mA injection current to the gain section is shown in Fig. 23B. Near-60 dB SMSRs were achieved within 1557–1585 nm, while greater than 50 dB SMSRs were seen across the entire range. The frequency noise of the laser was characterized with a commercial automated laser measurement system. Fig. 23C shows the power spectral density of the laser frequency fluctuation as a function of frequency, taken at ~4× threshold injection current (120 mA) and the wavelength of 1548 nm. $1/f$ (flicker) and $1/f^2$ (random walk) frequency noises are observed at the frequency range below 1 MHz. A white noise level

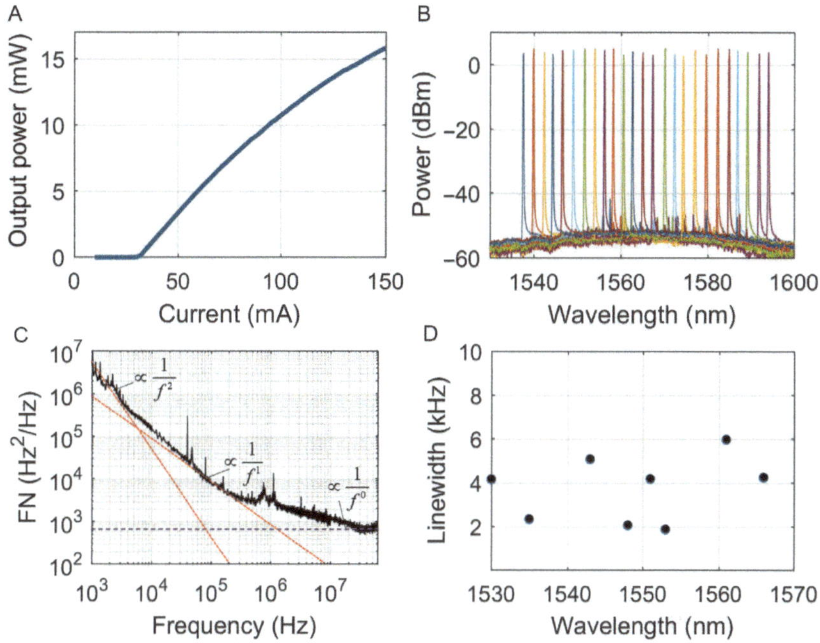

Fig. 23 (A) Laser output power vs injection current to the gain section. (B) Course tuning of the laser at 120 mA injection current to the gain section. (C) Measured frequency noise of the laser at 120 mA injection current to the gain section. The *red dash lines* indicate flicker (1/*f*) and random walk (1/*f*²) frequency noises, and the *blue dash line* indicates the white noise level. (D) Measured Lorentzian linewidths of the laser across the C band.

of $S_F(f) = 660\ \text{Hz}^2/\text{Hz}$ is seen at 30 MHz, corresponding to a Lorentzian linewidth of $\pi S_F(f) = 2.1$ kHz (Kikuchi and Okoshi, 1985). To the best of our knowledge, this is the narrowest linewidth among all fully integrated widely tunable lasers reported to date. Note that due to the interplay between optical negative feedback and effective cavity length enhancement in ring resonator mirrors, the best linewidths are always achieved slightly detuned from resonance. Nevertheless, even when operating at resonance peaks, which is common because it corresponds to the maximum output power and is easy to automate the ring tuning using an on-chip monitor photodetector, the linewidths were still narrower than 6 kHz across the C-band (Fig. 23D).

3.4 Mode-Locked Lasers

The mode-locked laser diode is a type of laser that, rather than emitting a continuous-wave output, emits a train of short optical pulses with repetition rates in the microwave frequency range. The short pulse, through the Fourier transform limit, results in a wide bandwidth in the optical frequency domain. Mode-locked lasers have a broad range of applications, including precision optical machining (Dausinger and Nolte, 2004), highly stable generation of microwave signals (Hall, 2000), broadband spectroscopy, optical analog to digital conversion (Callahan et al., 2012), and photonic radar (Ghelfi et al., 2014). These applications are typically conducted with large, expensive, and power hungry optically pumped bulk crystal or fiber lasers.

Mode-locked laser diodes offer the promise of semiconductor technology to the mode-locked laser: compact size, electrical pumping, and inexpensive manufacturing. Silicon, in particular, can bring the precision of CMOS manufacturing to the mode-locked laser, as well as superior passive waveguides when compared to InP-substrate lasers.

Fully integrated mode-locked laser diodes, meaning lasers that have an integrated mirror and a transition to passive waveguide to allow integration with other components inside a photonic circuit, typically lag behind their pure FP counterparts in terms of mode-locking performance, particularly peak output power. This is due to parasitic loss introduced by the integrated components (Davenport et al., 2018).

While InP-substrate mode-locked lasers have several decades of technological advancement over heterogeneous silicon/III–V lasers, the latter are rapidly catching up, as some of the best performing mode-locked laser diodes are now on silicon, specifically the microwave frequency linewidth (Wang et al., 2017).

Heterogeneous silicon mode-locked lasers often use the colliding pulse configuration, where the absorber is in the center of the cavity, as shown in Fig. 24, to avoid placing the mirror underneath the amplifier section. This is necessary for the use of loop mirrors for the broadband reflector, as the heterogeneous amplifier does not have as small bend radius as a passive silicon waveguide. Some implementations place a Bragg grating reflector in the silicon waveguide underneath the III–V mesa to implement a fundamentally mode-locked laser (Wang et al., 2017).

3.4.1 Absorber Design

The primary differentiator between a continuous-wave laser diode and a mode-locked laser diode is the inclusion of a saturable absorber. This is typically simply an electrically isolated segment of the amplifier mesa that is reverse biased. Because of high-speed gain saturation and recovery

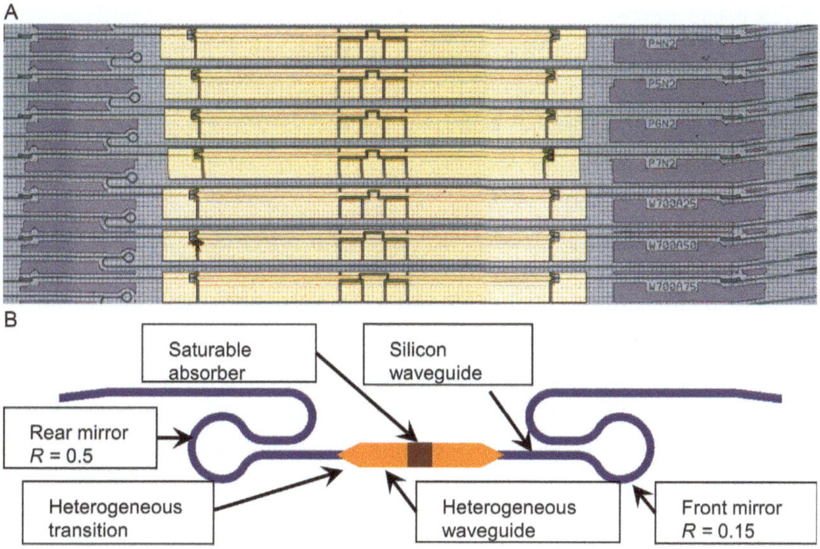

Fig. 24 Photo (A) and plan-view schematic (B) of a heterogeneous silicon mode-locked laser.

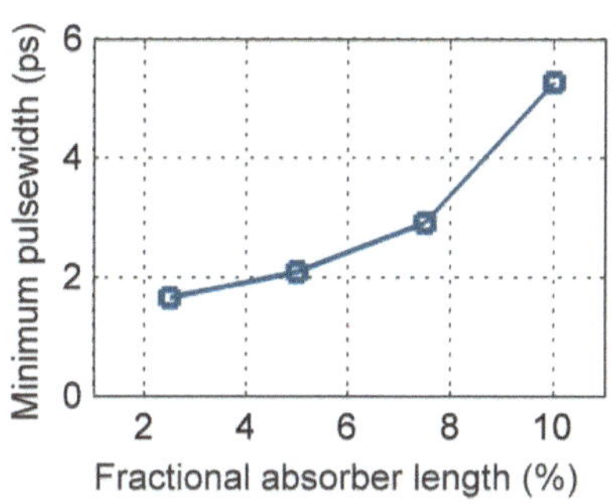

Fig. 25 Impact of absorber length on pulse width.

dynamics, this absorber section can cause the laser to pulsate at the round-trip frequency of the cavity and phase lock the optical modes, resulting in mode locking.

The length of the saturable absorber is an important parameter, as high reverse bias introduces a significant amount of loss inside the cavity, and so the absorber section should be as short as possible so that mode locking can be maintained at large negative reverse bias.

Fig. 25 shows the trend of minimum pulse width in a set of otherwise identical fully integrated mode-locked lasers (with the architecture shown in Fig. 24). The length of the amplifier section is 2 mm, and the x-axis of

Fig. 25 is the percentage of that value. The overall cavity length is 4 mm, to produce a 20 GHz pulse. Smaller saturable absorbers can operate at larger reverse bias, providing faster recovery time for the same insertion loss. This allows the formation of a shorter pulse.

3.4.2 Transition Design

The heterogeneous transition design ultimately has a high impact on mode-locked laser performance, as it does with any laser device that includes it. Loss, reflection, and higher order mode transmission must be minimized to as low levels as is practical (see Section 2.2). Because the mode-locked laser is a broadband device, it suffers from spectral impairments such as parasitic reflection and multimoding to a larger degree than single-wavelength devices, since these impairments reduce the available spectral bandwidth for the formation of a short pulse.

With this in mind, a special transition was designed for mode-locked lasers. This design uses a multisection taper to increase the adiabaticity of the device while still limiting the overall length to maintain low absorption loss. The concept is that as a twin waveguide system, the size of the upper waveguide (III–V mesa in this case) must be reduced in size to the termination to fully transform the mode into the silicon waveguide, but during this process, the second waveguide can be adjusted to maintain the same effective index throughout the structure, resulting in a more adiabatic transition. To accomplish this, while the III–V mesa tapers linearly from 4 μm at the end of the neck down to the termination at the taper tip, the underlying silicon waveguide is broken into three sections. The design is illustrated in Fig. 26. If the silicon waveguide is maintained at a fixed width, this results in a large change in effective index. Instead, during the initial section of the transition, in section I, there is no change in the silicon waveguide width. In section II, the silicon waveguide widens slightly to counteract the narrowing III–V mesa beginning to interact with the mode. In section III, the silicon waveguide width varies more rapidly to compensate for the now rapidly varying effective index, following along the effective index contour.

The resulting dimensions are as follows: the III–V taper is 30 μm long and is tapered linearly from 4 to 0.5 μm. The silicon tapers are each 10 μm. Taper I is 1.35 μm throughout. Taper II begins at 1.35 μm and tapers to 1.45 μm. Taper III tapers from 1.45 to 1.85 μm at the end of the p-type III–V mesa.

FDTD simulation suggests that this three-section conformal index transition should have 6 dB lower transmission into higher order modes than

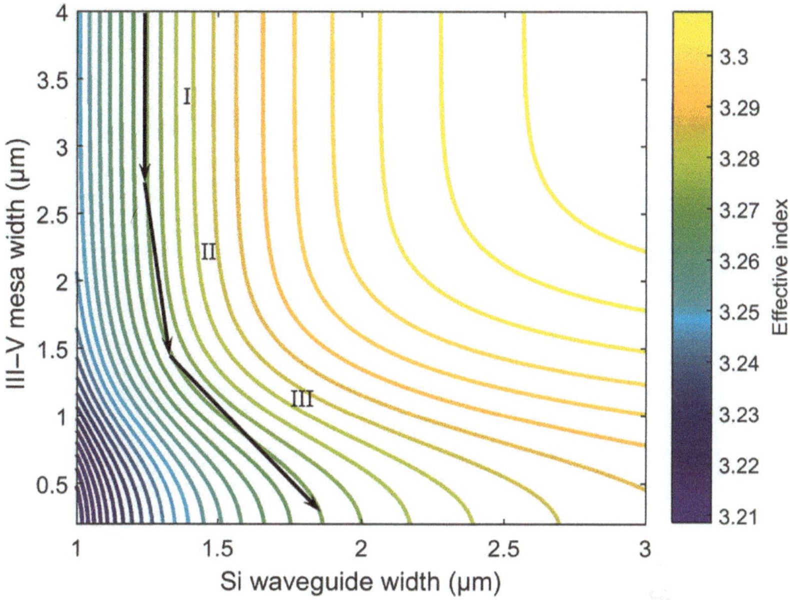

Fig. 26 Simulated effective index for a combination of III–V mesa widths and silicon waveguide widths.

the transition described in Section 2.2, −2. Because emission into higher order modes on these scales is difficult to measure directly, instead the pulse width of mode-locked lasers with the different transition designs is compared.

An experiment was devised in which two otherwise identical fully integrated mode-locked lasers were compared, one with a 15-μm-long three-section taper and the other with a 15-μm-long basic taper (like in Section 2.2). Fig. 27 shows that adopting the three-section taper improves the pulse width from 3.5 to 2.7 ps. Increasing the length to 30 μm reduces the pulse width further to 2.3 ps.

3.4.3 Cavity Design

The design of the laser cavity, by setting the reflections of the mirrors, is crucial for high output power and therefore high peak power. An advantage of fully integrated mode-locked lasers is that the reflections of the on-chip mirrors can be controlled during the design. High reflection in the rear mirror reduces the threshold current, and optimized reflection of the front mirror can increase output power. Too high reflection will prevent light from escaping from the cavity, and too low reflection will elevate the threshold current or even prevent lasing at all.

These mode-locked lasers use the architecture shown in Fig. 24 and have a 50% reflection rear mirror. The front mirror reflection was varied, and the resulting light–current characteristic was measured from a group of lasers. The results are shown in Fig. 28A, with the higher reflection

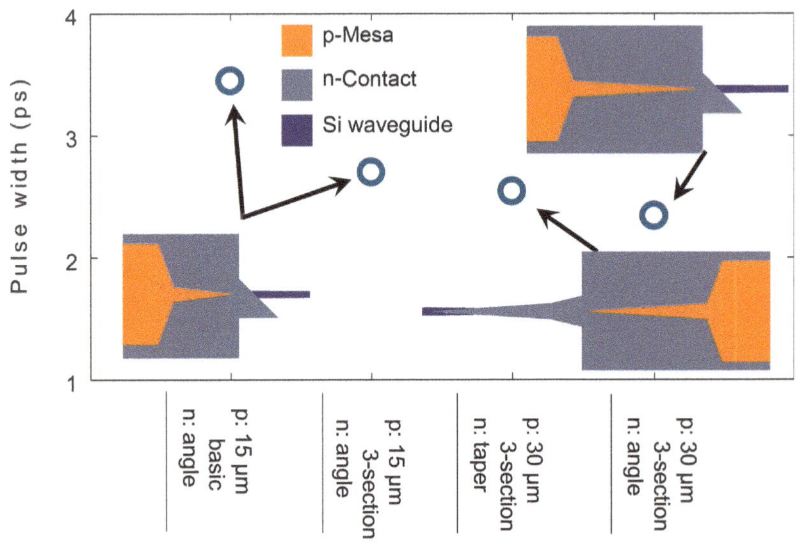

Fig. 27 Best recorded pulse width for four different types of heterogeneous transition.

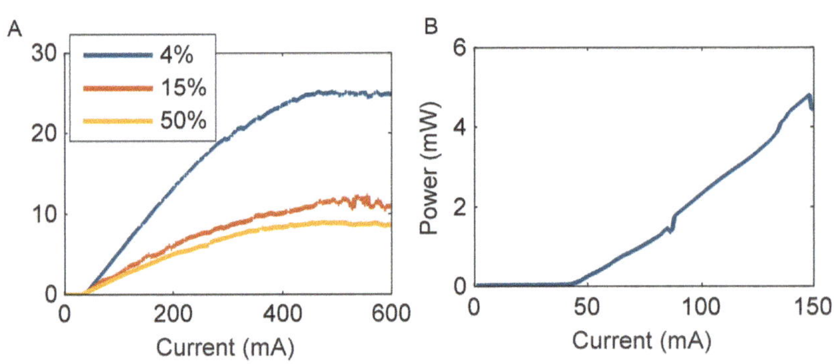

Fig. 28 Light–current characteristics for mode-locked lasers. (A) The impact of the front mirror reflectivity on slope efficiency and maximum output power, while the absorber is forward biased. (B) A detailed view around the operating condition that produced the shortest pulse from the final optimized design.

mirrors demonstrating lower output power, and a considerable advantage in power is displayed by the lowest reflection mirror. This experiment was performed with the saturable absorbers forward biased at the same voltage as the amplifier sections. Reverse biasing the absorber reduces the output power considerably, and the shortest pulse is typically not at the highest continuous-wave output power bias condition. A detailed view of the light–current characteristic around the bias condition for the shortest pulse is shown in Fig. 28A. This laser was biased at 98 mW on the gain section and −4.5 V on the absorber section when it produced its shortest pulse, and only outputs 1.78 mW of output power at this condition. However, it produces a 900-fs pulse, which corresponds to 98 mW of peak power at 20 GHz.

The trace from the autocorrelator is shown in Fig. 29A with its corresponding $sech^2$ fit. This is a fairly outstanding value for a mode-locked laser diode, particularly for a fully integrated variant (with on-chip passive waveguide and mirror), with only a handful of demonstrations of sub-1 ps fully integrated mode-locked laser diodes (Moskalenko et al., 2014; Parker et al., 2010). Fig. 29B shows the full span of the RF spectrum analyzer after receiving the pulse train on a high-speed photodetector, displaying the absence of large Q-switching peaks around 1 GHz, no trace of pulsation at the fundamental 10 GHz frequency, and a powerful narrow peak at the intended colliding-pulse operation at 20 GHz.

Fig. 29 The mode-locking performance of the final optimized design, showing (A) the measured autocorrelation trace for the 900-fs pulse and corresponding $sech^2$ fit, and (B) the RF spectrum for the device, showing a narrow signal at the intended 20 GHz colliding pulse frequency with little sign of *Q*-switching or mode locking at the cavity fundamental frequency.

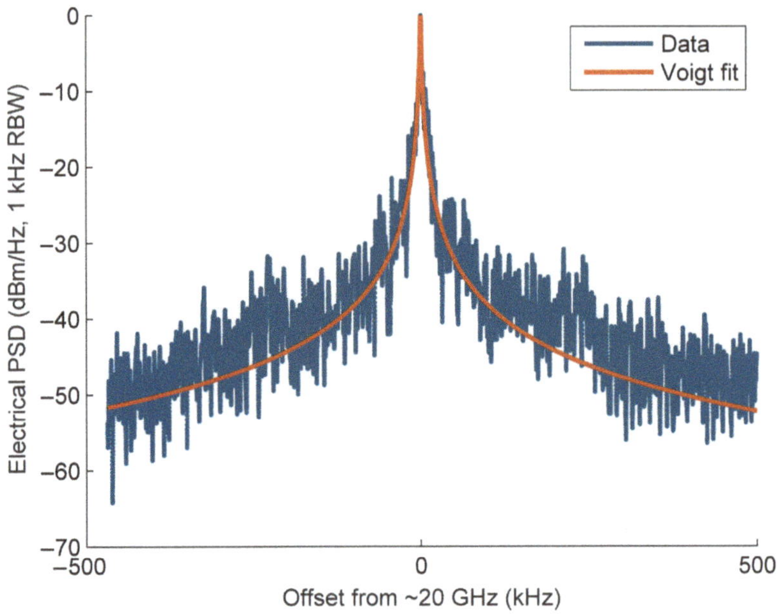

Fig. 30 Close-in view of the RF spectrum at the 20GHz peak showing the 1.1 kHz linewidth (*blue*) and a fitted Voigt spectrum.

This laser also demonstrated extremely low RF linewidth of 1.1 kHz (Fig. 30), which is an outstanding value which has currently only been achieved by silicon heterogeneous mode-locked lasers, as this result and the 900 Hz value from Wang et al. (2017). The superiority of silicon in the case of linewidth most likely stems from the lower loss passive cavities afforded by the silicon waveguide.

4. CONCLUSION

The combination of III–V materials and silicon with bonding overcomes the largest shortcoming of silicon photonics by allowing the function of light emission to be integrated with silicon photonic circuits. This type of integration allows wafer-scale fabrication of the III–V components, which is crucial for high-volume and low-cost manufacturing. The bonding is tolerant to misalignment, since the attachment of the unprocessed III–V chip only needs to cover the approximate area that the III–V devices will occupy. This is an important distinction when compared to flip-chip bonding completed SOAs into silicon devices, which requires submicron alignment accuracy.

This technology has been used to demonstrate a wide range of laser devices with high performance. Integration with multiple types of epitaxial wafer designs, which is normally complicated by regrowth, can be achieved with little trade-off by bonding separate epi wafer chips to the same die. Higher performance than native-substrate lasers is possible with heterogeneous integration, as the strengths of the InP platform and the silicon photonics platform can be combined. In particular, low-linewidth lasers, which are enhanced by long and low-loss passive waveguide cavities, benefit from this combination. State-of-the-art low-linewidth fully integrated mode-locked lasers, DFB lasers, and widely tunable lasers have all been demonstrated with heterogeneous silicon/III–V integration.

REFERENCES

Barbarin, Y., et al., 2005. Measurement of reflectivity of butt-joint active-passive interfaces in integrated extended cavity lasers. IEEE Photon. Technol. Lett. 17 (11), 2265–2267.

Ben Bakir, B., et al., 2010. In: Righini, G.C. (Ed.), Electrically driven hybrid Si/III-V lasers based on adiabatic mode transformers. Proceedings of SPIE, p. 77191F. http://proceedings.spiedigitallibrary.org/proceeding.aspx?doi=10.1117/12.858889.

Callahan, P.T., Dennis, M.L., Clark, T.R.C., 2012. Photonic analog-to-digital conversion. Opt. Express 30 (4), pp. 280–286.

Camacho-Aguilera, R.E., et al., 2012. An electrically pumped germanium laser. Opt. Express 20 (10), 11316.

Casey, H.C., Carter, P.L., 1984. Variation of intervalence band absorption with hole concentration in p-type InP. Appl. Phy. Lett. 44 (1), 82. Available at: http://scitation.aip.org/content/aip/journal/apl/44/1/10.1063/1.94561.

Cassidy, D.T., 1984. Technique for measurement of the gain spectra of semiconductor diode lasers. J. Appl. Phys. 56 (11), 3096. Available at: http://scitation.aip.org/content/aip/journal/jap/56/11/10.1063/1.333867. Accessed February 16, 2016.

Coldren, L., Corzine, S., Mashanovitch, M., 2012. Diode Lasers and Photonic Integrated Circuits, second ed. John Wiley and Sons.

Dausinger, F., Nolte, S., 2004. In: Dausinger, F., Lubatschowski, H., Lichtner, F. (Eds.), Femtosecond Technology for Technical and Medical Applications. Springer Berlin Heidelberg, Berlin, Heidelberg. Available at: http://link.springer.com/10.1007/b96440.

Davenport, M.L., et al., 2016. Heterogeneous silicon/III–V semiconductor optical amplifiers. IEEE J. Select. Top. Quant. Electr. 22 (6), 78–88. Available at: http://ieeexplore.ieee.org/document/7516667/.

Davenport, M.L., Liu, S., Bowers, J.E., 2018. Integrated heterogeneous silicon/III–V mode-locked lasers. Photon. Res. 6 (5), 468–478.

De Koninck, Y., Roelkens, G., Baets, R., 2013. Design of a hybrid III-V-on-silicon microlaser with resonant cavity mirrors. IEEE Photon. J. 5 (2), 2700413.

Doran, N.J., Wood, D., 1988. Nonlinear-optical loop mirror. Opt. Lett. 13 (1), 56–58.

Duan, G.H., et al., 2015. New advances on heterogeneous integration of III–V on silicon. J. Lightwave Technol. 33 (5), 976–983.

Fang, A., et al., 2006a. A continuous-wave hybrid AlGaInAs-silicon evanescent laser. IEEE Photon. 18 (10), 1143–1145. Available at: http://scholar.google.com/scholar?hl=en&btnG=Search&q=intitle:A+Continuous-Wave+Hybrid+AlGaInAs+?+Silicon+Evanescent+Laser#0. Accessed July 22, 2010.

Fang, A.W., et al., 2006b. Electrically pumped hybrid AlGaInAs-silicon evanescent laser. Opt. Express 14 (20), 9203.
Fang, A.W., et al., 2008. A distributed feedback silicon evanescent laser. Opt. Express 16 (7), 4413–4419.
Frey, B.J., Leviton, D.B., Madison, T.J., 2006. Temperature dependent refractive index of silicon and germanium. Proc. SPIE Int. Soc. Opti. Eng. 6273 II, 10. http://www.scopus.com/inward/record.url?eid=2-s2.0-33749642042&partnerID=tZOtx3y1%5Cnhttp://arxiv.org/abs/physics/0606168.
Fuchs, G., et al., 1992. Intervalence band absorption in strained and unstrained InGaAs multiple quantum well structures. Appl. Phys. Lett. 60 (2), 231–233.
Ghelfi, P., et al., 2014. A fully photonics-based coherent radar system. Nature 507 (7492), 341–345. Available at: http://www.nature.com/doifinder/10.1038/nature13078.
Gonzalez, L.P., Murray, J.M., Krishnamurthy, S., 2009. Wavelength dependence of two photon and free carrier absorptions in InP. Opt. Express 17 (11), 8741–8748.
Hall, J.L., 2000. Optical frequency measurement: 40 years of technology revolutions. IEEE. J. Sel. Top. Quant. Electr. 6 (6), 1136–1144.
Hawkins, A.R., et al., 1996. Silicon heterointerface photodetector. Appl. Phys. Lett. 68 (26), 3692–3694.
Heck, M.J.R., et al., 2013. Hybrid silicon photonic integrated circuit technology. J. Sel Top. Quant. Electr. 19 (4), 6100117. Available at: http://ieeexplore.ieee.org/lpdocs/epic03/wrapper.htm?arnumber=6387568%5Cnpapers3://publication/doi/10.1109/JSTQE.2012.2235413.
Hulme, J.C., et al., 2015. Fully integrated hybrid silicon two dimensional beam scanner. Opt. Express 23 (5), 5861. Available at: http://www.opticsinfobase.org/abstract.cfm?URI=oe-23-5-5861.
Hulme, J., et al., 2017. Fully integrated microwave frequency synthesizer on heterogeneous silicon-III/V. Opt. Express 25 (3), 2422.
Kazarinov, R.F., Henry, C.H., 1987. The relation of line narrowing and chirp reduction resulting from the coupling of a semiconductor laser to a passive resonator. IEEE J. Quant. Electr. 23 (9), 1401–1409.
Keyvaninia, S., Verstuyft, S., et al., 2013a. Heterogeneously integrated III-V/Si single mode lasers based on a MMI-ring configuration and triplet-ring reflectors. In: Proceedings of SPIE, vol. 8767, p. 87670N. http://proceedings.spiedigitallibrary.org/proceeding.aspx?doi=10.1117/12.2017327.
Keyvaninia, S., Muneeb, M., et al., 2013b. Ultra-thin DVS-BCB adhesive bonding of III–V wafers, dies and multiple dies to a patterned silicon-on-insulator substrate. Opt. Mater. Express 3 (1), 35. Available at: http://www.opticsinfobase.org/ome/abstract.cfm?URI=ome-3-1-35%5Cnhttp://www.opticsinfobase.org/DirectPDFAccess/63F72590-9FB2-DD17-5CBDB93B43265BF3_246989/ome-3-1-35.pdf?da=1&id=246989&seq=0&mobile=no%5Cnhttp://www.opticsinfobase.org/ome/abstract.cfm?uri=o.
Kikuchi, K., Okoshi, T., 1985. Estimation of linewidth enhancement factor of AlGaAs lasers by correlation measurement between FM and AM noises. IEEE J. Quantum Electron. 21 (6), 669–673.
Kobayashi, N., et al., 2015. Silicon photonic hybrid ring-filter external cavity wavelength tunable lasers. J. Lightwave Technol. 33 (6), 1241–1246.
Koch, B.R., et al., 2013. In: Integrated silicon photonic laser sources for telecom and datacom. Optical Fiber Communication Conference/National Fiber Optic Engineers Conference 2013. p. PDP5C.8.
Komljenovic, T., et al., 2016. Heterogeneous silicon photonic integrated circuits. J. Lightwave Technol. 34 (1), 20–35.
Komljenovic, T., et al., 2017. Widely-tunable ring-resonator semiconductor lasers. Appl. Sci. 7 (7), 732.

Kreissl, J., et al., 2000. In: Hydrogen passivation in InP:Zn resulting from reactive ion etching during laser stripe formation. Conference Proceedings. 2000 International Conference on Indium Phosphide and Related Materials (Cat. No.00CH37107). IEEE, pp. 142–145. Available at: http://ieeexplore.ieee.org/document/850252/.

Kurczveil, G., et al., 2016. Robust hybrid quantum dot laser for integrated silicon photonics. Opt. Express 24 (14), 16167–16174.

Liang, D., Fiorentino, M., et al., 2011a. Low threshold electrically-pumped hybrid silicon microring lasers. IEEE. J. Sel. Top. Quant. Electr. 17 (6), 1528–1533.

Liang, D., Chapman, D.C., et al., 2011b. Uniformity study of wafer-scale InP-to-silicon hybrid integration. Appl. Phys. A Mater. Sci. Process. 103 (1), 213–218.

Liang, D., et al., 2016. Integrated finely tunable microring laser on silicon. Nat. Photonics 10 (11), 719–722.

Liu, B., Shakouri, A., Bowers, J.E., 2001. Passive microring-resonator-coupled lasers. Appl. Phys. Lett. 79 (22), 3561–3563.

Liu, B., Shakouri, A., Bowers, J.E., 2002. Wide tunable double ring resonator coupled lasers. IEEE Photon. Technol. Lett. 14 (5), 600–602.

Matsuo, S., et al., 2014. Directly modulated buried heterostructure DFB laser on SiO_2/Si substrate fabricated by regrowth of InP using bonded active layer. Opt. Express 22 (10), 12139. Available at: https://www.osapublishing.org/oe/abstract.cfm?uri=oe-22-10-12139.

Moskalenko, V., et al., 2014. Record bandwidth and sub-picosecond pulses from a monolithically integrated mode-locked quantum well ring laser. Opt. Express 22 (23), 28865. Available at: https://www.osapublishing.org/oe/abstract.cfm?uri=oe-22-23-28865.

Moutanabbir, O., Gösele, U., 2010. Heterogeneous integration of compound semiconductors. Annu. Rev. Mat. Res. 40 (1), 469–500.

Nakamura, M., et al., 1973. Optically pumped GaAs surface laser with corrugation feedback. Appl. Phys. Lett. 22 (10), 515–516.

Parker, J.S., et al., 2010. Integrated 30 GHz passive ring mode-locked laser with gain flattening filter. Conference Digest—IEEE International Semiconductor Laser Conference 1 (805), 3–4.

Roelkens, G., et al., 2015. III-V-on-Silicon photonic devices for optical communication and sensing. Photonics 2, 969–1004.

Santis, C.T., et al., 2014. High-coherence semiconductor lasers based on integral high-Q resonators in hybrid Si/III-V platforms. Proc. Natl. Acad. Sci. U.S.A. 111 (8), 2879–2884. Available at: http://www.pnas.org/cgi/doi/10.1073/pnas.1400184111.

Santis, C.T., Vilenchik, Y., Yariv, A., 2015. Sub-kHz Quantum Linewidth Semiconductor Laser on Silicon Chip. CLEO 2015 Post Deadline Paper Digest. vol. 1, pp. 1–2.

Spott, A., et al., 2017. Heterogeneous integration for mid-infrared silicon photonics. IEEE J. Select. Top. Quant. Electr. 23 (6), 1–10.

Srinivasan, S., et al., 2011. Design of phase-shifted hybrid silicon distributed feedback lasers. Opt. Express 19 (10), 9255.

Srinivasan, S., et al., 2015. Coupled-ring-resonator-mirror-based heterogeneous III-V silicon tunable laser. IEEE Photon. J. 7 (3), 1–8.

Sun, X., et al., 2013. Tunable silicon Fabry-Perot comb filters formed by Sagnac loop mirrors. Opt. Lett. 38 (4), 567–569. Available at: http://www.ncbi.nlm.nih.gov/pubmed/23455138.

Tran, M.A., et al., 2017. Integrated optical driver for interferometric optical gyroscopes. Opt. Express 4 (4), 3826–3840.

Van Campenhout, J., et al., 2007. Electrically pumped InP-based microdisk lasers integrated with a nanophotonic silicon-on-insulator waveguide circuit. Opt. Express 15 (11), 6744. Available at: https://www.osapublishing.org/oe/abstract.cfm?uri=oe-15-11-6744.

Wang, Z., et al., 2017. A III-V-on-Si ultra-dense comb laser. Light Sci. Appl. 6 (5), 16260. Available at: https://doi.org/10.1038/lsa.2016.260.

Wang, Y., et al., 2018. High-power photodiodes with 65 GHz bandwidth heterogeneously integrated onto silicon-on-insulator nano-waveguides. IEEE J. Select. Top. Quant. Electr. 24 (2), 1–6.

Xie, X., et al., 2015. Heterogeneously integrated waveguide-coupled photodiodes on SOI With 12 dBm output power at 40 GHz. In: Optical Fiber Communication Conference Post Deadline Papers, p. Th5B.7.

Zhang, C., et al., 2014. Low threshold and high speed short cavity distributed feedback hybrid silicon lasers. Opt. Express 22 (9), 10202. Available at: https://www.osapublishing.org/oe/abstract.cfm?uri=oe-22-9-10202.

Zhang, C., et al., 2016. $8 \times 8 \times 40$ Gbps fully integrated silicon photonic network on chip. Optica 3 (7), 785.

CHAPTER SEVEN

InP Photonic Integrated Circuits on Silicon

Jos J.G.M. van der Tol[1], Yuqing Jiao, Kevin A. Williams
Photonic Integration Group, Eindhoven University of Technology, Eindhoven, The Netherlands
[1]Corresponding author: e-mail address: j.j.g.m.v.d.tol@tue.nl

Contents

Semiconductors and Semimetals, Volume 99
ISSN 0080-8784
https://doi.org/10.1016/bs.semsem.2018.08.005

IMOS (InP membrane on silicon) is a platform-based approach to create indium phosphide (InP) nanophotonic integrated circuits. It uses monolithic integration of all photonic functions in the InP layers to leverage the most efficient optoelectronic processes and to remove optical interfaces between separately grown wafers. Using an optoelectronic material for the photonic components provides a powerful route to nanoscale miniaturization and circuit-level performance enhancements.

In this chapter, we review the progress in the passive and active InP building blocks (BBs) and highlight routes to miniaturization and performance scaling. We address the interfaces between BBs, which are critical to a powerful platform approach. The platform approach enables customization by designers to create new components and circuits. The process design kit (PDK) methodology is applied to provide a route to circuit design abstraction.

1. NANOPHOTONIC INTEGRATION PLATFORMS

Heterogeneous integration of InP on Silicon has been proposed for large-scale manufacturability (Abbasi et al., 2016; Jain et al., 2011), integration of photonics with electronics (Arakawa et al., 2013; Tol et al., 2018), and to generate and amplify light for Silicon-based circuits. Miniaturizing InP optoelectronic devices and circuits offers the additional prospect of reduced energy consumption which has particular benefits in ensuring sustained bandwidth and performance increases for photonic ICs. The potential advantages are most strikingly evident for data connections, data centers, and the Internet, but energy-efficient photonics is expected to impact a broader range of applications, like the Internet of things photonic sensors for, e.g., precision gesture recognition, proximity sensing, THz analysis, and free-space data connectivity (e.g., LiFi).

Nanophotonic components and circuits are created in a (sub)micron-scale thickness membrane which is mechanically supported by a Silicon carrier wafer. A broad range of techniques have been proposed in the literature for combining InP with Silicon, and in many, the Silicon plays a role in the optical waveguiding. The term *Silicon Photonic* is often used as a catch-all for such approaches. In this chapter, we shall not use the term, focussing instead on the methods of integration of nanophotonic InP BBs which are optically isolated from the Silicon.

The main approaches to enable nanophotonic circuit integration of active and passive components through high-index-contrast material combinations include:

- Molecular bonding of InP epitaxial material on top of silicon-on-insulator (SOI) waveguides which are subsequently processed to define gain sections for lasers (Davenport et al., 2016) through evanescent coupling from the underlying Silicon waveguides.
- Adhesion bonding of InP epitaxial material including optical waveguiding layers (Duan et al., 2014) to enable light from Silicon waveguides to couple up into postprocessed amplifiers.
- Nanophotonic membranes, directly bonded to Si/SiO_2, with lateral carrier injection to enable size reduction of active components and the use of a Silicon carrier wafer for manufacturing (Matsuo et al., 2015).
- Selective area epitaxial growth and subsequent etching to create InP nanobeams on Silicon have been proposed and demonstrated for optically pumped lasers (Wang et al., 2015). Considerable research will be required to develop methods for electronic pumping.
- Transfer printing (Corbett et al., 2017) has been developed over the years, including epitaxial lift-off and wafer bonding with substrate removal. Recently a new technique (Zhang et al., 2018), also called transfer printing, has been introduced which allows manipulation of small and thin materials along with devices on a massively parallel scale, allowing micron-scale placement accuracies.
- IMOS—an InP membrane on Silicon platform which enables a full suite of nanophotonic components to be monolithically integrated (van der Tol et al., 2014).

The above methods leverage nanophotonic waveguides with an order of magnitude lower cross-sectional areas when compared to commercially available optoelectronic devices, but so far component areas have not been significantly reduced without incurring considerable performance tradeoffs. The monolithic integration inherent to the IMOS platform avoids assembly-dependent transitions, leverages the most efficient optoelectronic BBs, and provides a rich flexibility in BB-level optimization and interconnection.

1.1 Indium Phosphide Membrane on Silicon

InP membranes offer a powerful platform approach to leverage active and passive device miniaturization to create smaller circuits within the same wafer-scale processes. The IMOS platform offers a clear route to footprint reduction, energy efficiency, and high bandwidth for lasers, modulators, and detectors.

Lasers with energy efficiency are achieved through cavity size minimization and high-reflectivity cavities with high-contrast reflectors. The low loss cavity concepts deployed for vertical cavity lasers in low-energy data communications, consumer electronics, and gesture and face recognition may also be anticipated through the combination of high-efficiency in-plane gain media and high-reflectivity photonic crystal (PhC) cavities (Fujii et al., 2018; Inoue et al., 2016). A route to waveguide-coupled nanopillar light sources is also feasible (Dolores-Calzadilla et al., 2017). Nanophotonic filtering combined with modulators ensures compatibility with dense wavelength division multiplexing and complex modulation formats (Kish et al., 2018) for higher functionality and lower footprint, supporting data, and sensing circuits. By combining active and passive components in the same monolithic technology, the interfaces become highly controlled, removing optical parasitic impairments such as spurious reflections (Barbarin et al., 2005) and intracavity losses (Kurczveil et al., 2013) which can be seen in hybrid integration schemes.

Amplitude modulators enable data transmission which is unimpeded by nonlinear laser dynamics. Quantum-confined Stark effect (QCSE) (Shen et al., 2016a) and Franz–Keldysh effect (FKE) modulators (Rouvalis et al., 2012) are now key enablers for intermediate transmission distances in data communications, but can suffer from restricted optical bandwidths of the order 10–20 nm. Electro-absorption modulators (EAMs) using Fermi-level variation in graphene on an SOI waveguide show optical bandwidths in excess of 80 nm (Xie et al., 2014), but with low extinction ratio and limited modulation frequency due to high contact resistance to graphene. Poorer extinction ratios ultimately lead to poorer energy performance through an increased optical power budget. The IMOS platform can use precision doping in nanophotonic waveguides for a new form of modulator design: highly doped n-InGaAs shows large changes in the absorption when the electron concentration is varied (Ishibashi et al., 1997; Pello et al., 2013).

Phase modulation through the Kerr effect in an integrated Mach–Zehnder interferometer (MZI) has become the workhorse for telecommunications, but common implementations require several hundreds of microns (InP) or even millimeters (Silicon) of phase modulator length for linear, high extinction modulation. Promising new highly efficient electro-optic (EO) polymers (Pello et al., 2012) benefit from high confinement slot waveguide techniques which are readily implemented in a high confinement platform (Dolores-Calzadilla et al., 2014).

Detector technology is assessed in terms of speed, responsivity, saturation properties, dark current, and noise performance. InGaAs photodiodes offer the most sensitive detection and, when combined with precision doping and high resolution lithography, provide a route to creating high optical power and high-speed operation (Shen et al., 2016a). In the past decade the UTC-PD and its variants have been shown to exhibit very high bandwidth and high saturation power in InP-based PICs (Ishibashi et al., 1997; Rouvalis et al., 2012; Xie et al., 2014). Unlike in p-i-n PDs, only electrons act as the active carrier in the depletion region of UTC-PDs leading to a reduced transit time and reduced space charge blocking.

Wavelength- (Pello et al., 2013), polarization- (Pello et al., 2012), and surface grating- (Dolores-Calzadilla et al., 2014) based passive components for lightwave manipulation leverage the low-topology, high-precision photonic structures which can now be made together on the same active–passive integration platforms to provide one of the most comprehensive ranges of components on a platform technology.

1.2 Chapter Outline

In this chapter, we present the IMOS platform. Section 2 outlines the technological approach. Section 3 reviews the passive devices which are created using the platform and Section 4 highlights the active devices including high-speed detectors and integrated lasers and amplifiers. Section 5 provides an outlook in terms of opportunities for process-design-kit-supported production, routes to scaling up for manufacturing, and heterogeneous integration with other semiconductor technologies such as microelectronics.

2. TECHNOLOGY

The technologies developed for InP-substrate PICs (Smit et al., 2014) have matured over the last decades and provide the technological foundation for the IMOS platform. However, several key processes need to be further optimized to accommodate the higher optical confinement (see the optical mode profiles, Fig. 6) in the membrane waveguides. This section discusses the key technological challenges and innovations for InP membrane photonics.

2.1 Layer Structure

High optical confinement is achieved by removing the InP substrate to create a photonic membrane. This step is performed after the InP epitaxial wafer has been adhesion bonded to a silicon carrier with a low-index benzocyclobutene (BCB) polymer layer. The same 300-nm-thick passive waveguiding layer connects all the active and passive components throughout the membrane. Active structures with doped layers are defined in close proximity to the waveguiding layer and coupled to the latter evanescently. A schematic illustration of the layer structure is shown in Fig. 1 (van Engelen et al., 2017b). The cross-sectional view shows the possibility to integrate photodetectors (PDs), passive waveguide devices, and phase modulators exploiting EO polymers in slot waveguides and optical amplifiers together with the mode-adaption elements for interconnecting devices. The silicon carrier acts as a mechanical support for the membrane circuits and can also be used as a heat sink. This approach is expected to enable future integration with other wafer-scale technologies such as Silicon electronics. The devices shown in the figure will be discussed in the following sections.

2.2 Double-Sided Processing

The InP membrane allows for waveguide topologies to be created on both sides of the membrane. Waveguides (and other structures) may be created prior to and subsequent to wafer bonding. Such double-sided processing technology opens an extra degree of freedom in the design and optimization of membrane devices (Shen et al., 2016b). A schematic process flow is shown in Fig. 2. The topology created by the processing steps before bonding are covered and planarized by the bonding adhesive. After bonding and substrate removal, further process steps can be performed on the clean and

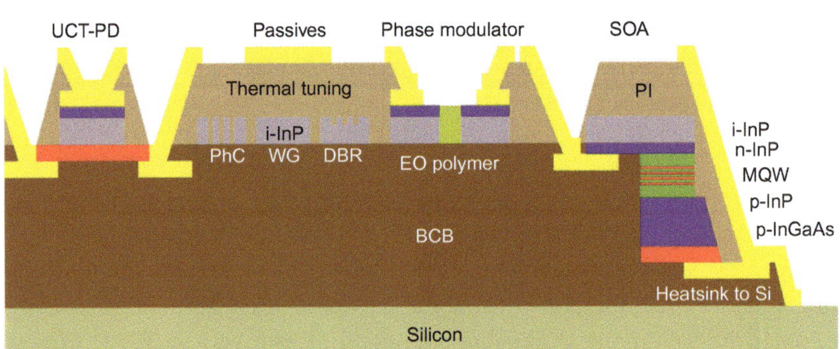

Fig. 1 Schematic illustration of the InP membrane layer structure (van Engelen et al., 2017b).

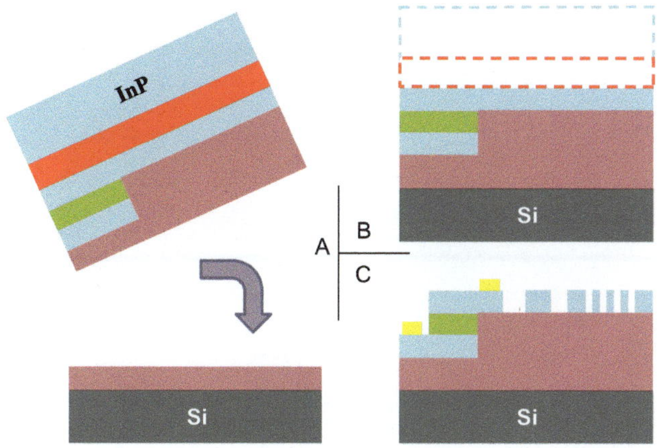

Fig. 2 Schematic process flow to show double-sided topology for InP membrane. (A) Processing before bonding, (B) bonding and substrate removal, and (C) processing after bonding.

planar back surface of the membrane. The structures created after bonding are aligned to the previously created, top-side structures, using predefined, through-wafer markers.

2.3 Bonding

The bonding process is the key enabler to the double-sided InP membrane platform. Adhesive bonding with BCB polymer is highly suitable for this purpose. The refractive index of BCB at 1.55 μm wavelength is 1.54, close to silicon oxide, making it suitable as a low-index optical buffer. The BCB is widely used as a planarization polymer for complex topologies with large step heights (Burdeaux et al., 1990) as well as bonding adhesive with high flexibility and robustness (Niklaus et al., 2006). The deposition of BCB is based on spin coating. Fig. 3 shows 3-in. InP to 3-in. Silicon full wafer bonding. After removing the InP substrate, a uniform 300-nm-thick InP membrane remains on top of the BCB/SiO_2/Si-bonding stack. Such a wafer is ready for further lithography and etching processes.

2.4 Lithography

Nanophotonic waveguide definition requires lithography with an order of magnitude higher resolution than micron-scale, substrate-based InP PICs. Target critical dimensions are the waveguide width (400 nm), distributed Bragg reflector (DBR) grating half-pitch (150 nm), and PhC reflector hole diameter (100 nm).

Fig. 3 3-in. InP to Silicon wafer bonding. Top surface is a uniform 300–nm-thick InP membrane. *Image by Y. Jiao. The picture was taken at an angle, and the* shallow white stripe *is a reflection from the ceiling lighting.*

Electron beam lithography (EBL) with ZEP resist is used for the patterning. Baking at a temperature close to the reflow temperature can significantly reduce the sidewall roughness, which is critical for low loss waveguides. However, fine patterns made from ZEP resist may suffer from deformation during the baking, as shown in Fig. 4A, as well as from roughness during dry etching. Deformations are due to the material softness, which results in a low resistance to thermal and plasma treatment. A novel resist mixture has been developed (Jiao et al., 2014). The mixture contains C_{60} nanoparticles which can fill in the voids among ZEP molecules (Ishii and Shigehara, 2009). The mixture has been proven to have enhanced thermal stability and plasma resilience. As a result, the accuracy and roughness of the patterns can be significantly improved, as can be seen in Fig. 4B, leading to the avoidance of pattern collapse on the more exposed grating elements.

The writing time for EBL is expected to become a limiting factor as circuit complexity and wafer size increase. DUV scanner lithography is a promising solution in that case. It has already been applied in generic InP foundry processing (Augustin et al., 2018). Scanner exposure processes are now being developed for the InP photonic membrane.

2.5 Etching

Methane (CH_4)-based dry etching is used for InP optical waveguide realization. Excess polymer generation does however induce roughness on the

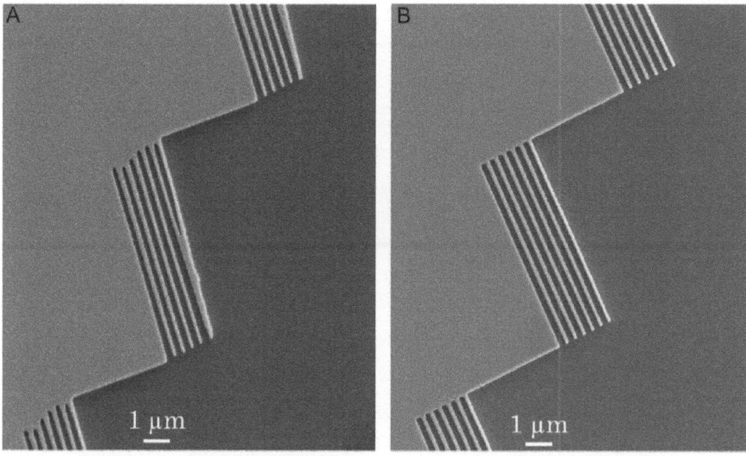

Fig. 4 DBR grating pattern made from (A) ZEP resist and (B) C60/ZEP mixture (Jiao et al., 2014).

etched surface. On the other hand, polymer generation is a powerful means to protect the etched surface from the ion-induced physical damage. Thus a delicate balance is required between polymerization and physical erosion. A dry etching process based on the $CH_4/H_2/Ar$ combination has been developed (Jiao et al., 2015). The ratio of the gases is optimized to achieve this balance during the entire etching process of the 300-nm-thick membrane, without the need for O_2 interruption to remove organic etching products from the surfaces (Karouta, 2014). An etched waveguide structure with smooth surface is shown in Fig. 5. Resulting waveguide losses as low as 2.5 dB/cm have been measured (Mejia, 2018; Zhu et al., 2017).

2.6 Electrical Contact

In InP photonic membranes, the high optical confinement shrinks significantly the dimension of devices. One of the consequences is the much closer distance between the optical mode and the optically lossy metal contacts. A Ge/Ag metal stack is developed as an n-contact with low-optical-loss and low-contact resistance (Shen et al., 2015). Annealing of metals typically causes spiking of metals into the semiconductor and results in additional optical loss. It has been discovered that in combination with heavily n-doped quaternary and ternary layers, Ge/Ag contact resistance can be sufficiently low (below $10^{-6}\,\Omega\,cm^2$) without the need for rapid thermal annealing (Shen et al., 2016c). Additionally, sixfold reductions of optical loss have been verified with Ge/Ag, when compared with commonly used Ti/Pt/Au metal stacks.

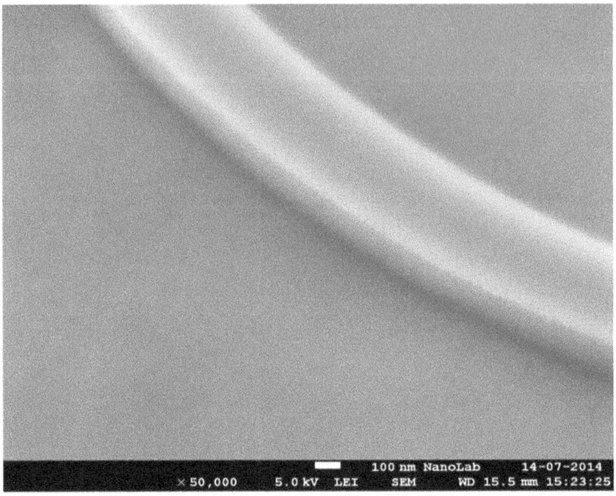

Fig. 5 SEM picture of a curved waveguide etched by the optimized $CH_4/H_2/Ar$ process (Jiao et al., 2015).

2.7 Discussion

The technology development in the InP photonic membrane platform has enabled wafer-scale nanophotonics with monolithic active–passive integration. Full integration of active (including intrinsic lasers) and passive BBs can be achieved. Devices can be connected by low-optical-loss nanophotonic waveguides to build complex functional circuits. In the following sections selected active and passive devices are presented.

3. PASSIVE DEVICES

The passive waveguide enables interconnection between the optoelectronics devices and is therefore at the heart of the membrane photonic platform. All functional devices are connected through a 300-nm-thick undoped InP layer in which 400-nm-wide rectangular waveguides are etched. These dimensions are chosen to provide effectively single-mode operation, while obtaining high confinement of the guided modes. The intensity profiles of the TE- and the TM-like waveguide modes are given in Fig. 6. Using the optimized technologies described in the previous section, high-quality waveguides have been demonstrated. Record low propagation loss for InP membrane waveguides of 2.5 dB/cm has been obtained (Jiao et al., 2014, 2015).

Based on these waveguides, a series of passive devices have been developed, which will be described in the following sections.

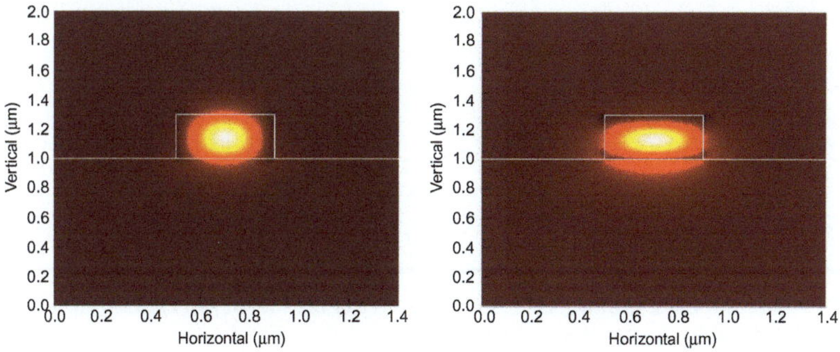

Fig. 6 Intensity profiles of the TE-like (*left*) and TM-like (*right*) guided modes of the standard InP membrane waveguide.

3.1 Wavelength Demultiplexers

In order to be a relevant integration platform for telecom applications, it is essential to include wavelength demultiplexing into the function set. Arrayed waveguide grating (AWG) offers a powerful way to filter and de/multiplex light. The mode of operation is based on the wavelength-dependent propagation through a set of parallel, usually curved, waveguides of different lengths. A specific problem with making such a structure on a membrane is the enhanced sensitivity to phase errors, requiring improved critical dimension uniformity for high-contrast waveguides. Nevertheless an AWG has been made in an InP membrane. An AWG of 0.2 mm^2 surface area showed a cross talk level of 10 dB and a loss of also 10 dB (Tol et al., 2018).

In order to relax the critical dimension uniformity requirement for high-contrast waveguides, another option has been developed: the planar concave grating (PCG). The principle is based on a reflective grating. This concept is not popular in low-contrast photonic integration, because it is difficult to etch the very high aspect ratio gratings perfectly perpendicular to the waveguide. However, in a high-contrast membrane this is much less of a problem, because of the high numerical aperture of the waveguide. The PCG in the InP membrane is based on a design from Pogoretskiy et al. (2017a). It is similar in size as the AWG mentioned before, but showed a much improved performance: a cross talk of 25 dB and a loss of 4 dB (Niklaus et al., 2006). Fig. 7 shows the device structure and the measured output spectra. Of note is here that the spectral response could be tuned with a thermal heater, allowing easy alignment of the device with the ISU wavelength grid (Pogoretskiy et al., 2017b).

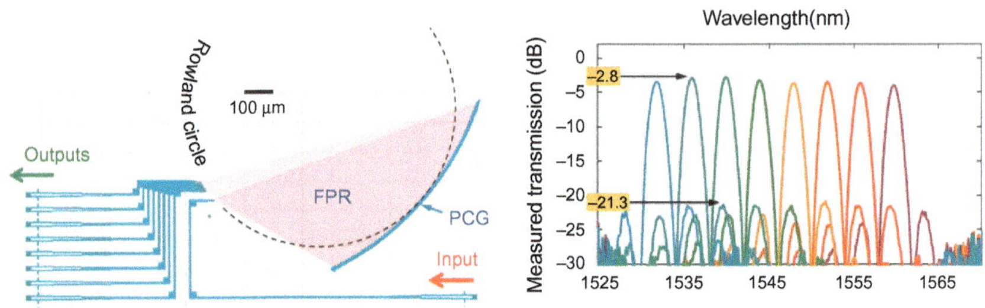

Fig. 7 Planar concave grating. *Left*: Structure of the device. *Right*: Experimental spectra from eight outputs (Pello et al., 2013).

3.2 Couplers

A basic functionality in photonic circuits is the splitting and combining of propagating lightwaves. This is used in Mach–Zehnder interferometers. These couplers are mostly based on multimode interference (MMI) devices, because of their small size, tolerance against deviations, wavelength insensitivity, and polarization independence. For the couplers on the InP membrane the well-known concepts from low-contrast photonic integration can be used, but the devices will be an order of magnitude smaller. There is however one serious problem arising from this: due to the high-index contrast, the MMIs are very prone to reflections. Reflection levels of several percent can be found if no design precautions are taken, which is quite detrimental when using them in interferometric structures. The concept from Pogoretskiy et al. (2017b) is used to reduce reflections as much as possible: making an open structure, allowing light that would otherwise reflect of damaging interfaces, to escape into a surrounding planar waveguide.

Fig. 8 shows the operation of this coupler. The worst-case condition w.r.t. reflection is the combination of two out-of-phase optical waves. In that case no light enters the output waveguide, so all the energy has to escape along a different path. For a traditional rectangular-shaped MMI there is a strong coupling to the reflection path, and up to 10% (10 dB) can return into the input waveguides. Tilting the interfaces next to the output waveguide reduces the reflection levels significantly to 0.1% (30 dB), but the open structure clearly performs best, with reflections dropping to 0.04% (38 dB) (Shen, 2016).

3.3 Polarization Converters

A full set of functions for a photonic integration platform must include the possibility to manipulate the state of polarization of the light. Although the modes in the waveguide are well defined and stable with a TE-like or

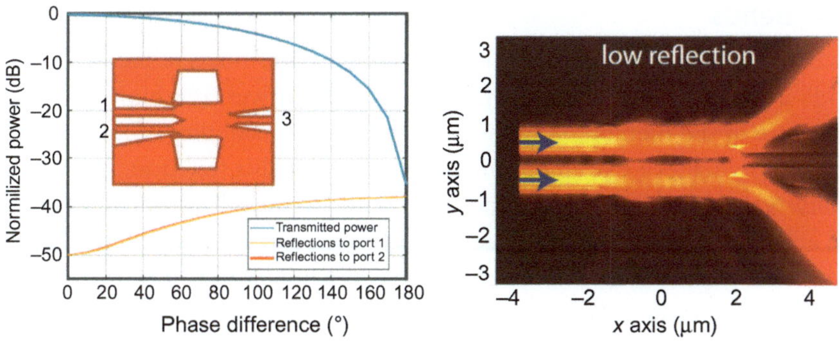

Fig. 8 Low reflection 2-to-1 MMI-coupler (*inset left*). *Left*: Transmission to the output (*blue*) and the reflections to the upper (*yellow*) and lower (*red*) input ports as a function of the phase difference of the input modes. *Right*: The electric field TE mode distribution for out-of-phase inputs (Mejia, 2018).

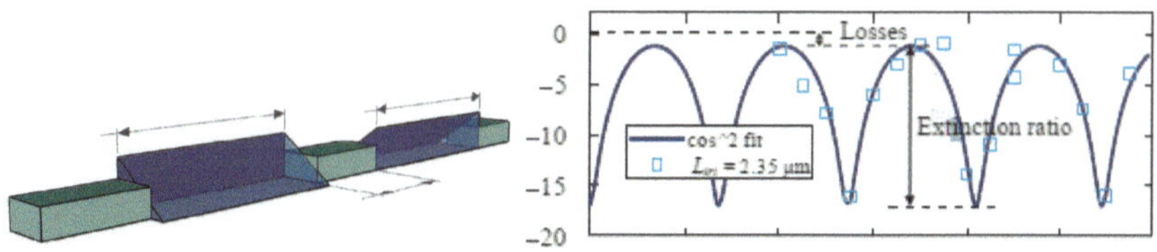

Fig. 9 *Left*: Polarization converter consisting of *two triangular sections*. *Right*: Transmission (in dB) of the TE-mode as a function of the waveguide length between the sections. The interference pattern reveals the losses and the polarization extinction ratio (Pello et al., 2012).

a TM-like polarization, this does not hold for the world outside the integrated circuit; light from a standard single-mode fiber can have a varying polarization state. To accommodate this, and to allow the functional use of polarization in, e.g., polarization multiplexing, a basic BB is needed: a polarization converter. This device will couple (fully or partly) the TE- and TM-like modes. To achieve this specially designed waveguide cross-sections are used, which carry modes with tilted orientation with respect to the vertical and horizontal directions on the chip. One such device is shown in Fig. 9 (Pello et al., 2012). It consists of two triangular sections, connected to standard rectangular membrane waveguides. Each of the sections results in a rotation of the polarization by 45 degrees, so in total a full conversion from TE to TM (or vice versa) is obtained. The triangular shape is created with a self-limiting wet etch. The experimental polarization converter showed >99% conversion and <1 dB loss, with a length of 5 μm.

3.4 Bends

S-bends can be realized in an InP membrane with much reduced bending radii as compared to generic InP platforms. However, the high numerical aperture of the membrane waveguides offers the possibility of using corner mirrors to achieve extremely small footprints for this function, in order to allow very dense photonic circuits. In InP-substrate PICs this idea is not so effective, because the losses are highly sensitive to a vertical tilt of the mirror, but in membranes this sensitivity is much reduced due to the high numerical aperture. To optimize for low loss and low reflections a cavity is designed around the mirror (see Fig. 10; Jiao et al., 2016a).

In Fig. 10 experimental results for both the corner mirrors and S-bends are shown. Lowest losses are obtained with relatively large radius (>5 μm) S-bends: around 0.04 dB/90 degrees. However, for the small footprint bends the corner mirror outperforms the S-bends on both losses (0.13 dB/90 degrees) and size (effective radius 0.96 μm).

3.5 Gratings

A major challenge for nanophotonic waveguides is the mismatch in mode size with respect to optical fibers. Direct butt coupling of a fiber and a membrane waveguide is very inefficient. A 3D inverted taper can be used to accommodate this (Pu et al., 2010), but the length of such a structure is considerable, up to several millimeters. Therefore, another technique is often used for input and output of light: a diffractive grating. If the phase-matching condition between an incoming field, the mode propagating in the waveguide, and the grating is fulfilled, a coupling will be obtained. To optimize this coupling, width, period, filling factor, strength, and length of the grating can be adjusted (Zhu et al., 2017). Additional measures are using apodization, so that the diffracted output field matches more closely to the mode in the fiber, and using a back-side mirror on the membrane (Higuera-Rodriguez et al., 2015). This mirror increases the efficiency, since the part of the light that travels through the membrane can in this way be added to the fiber–waveguide coupling. Without a back mirror the IMOS-grating losses are 5 dB, but with a silver mirror implemented 2.7 dB is measured. A grating structure including this mirror is depicted in Fig. 11. The metal reflector additionally serves to optically isolate the circuit from the underlying handling wafer. Another trick is using subwavelength structures within each grating period, in order to obtain the required grating strength, while reducing reflection (Millan-Mejia et al., 2016). The grating apertures

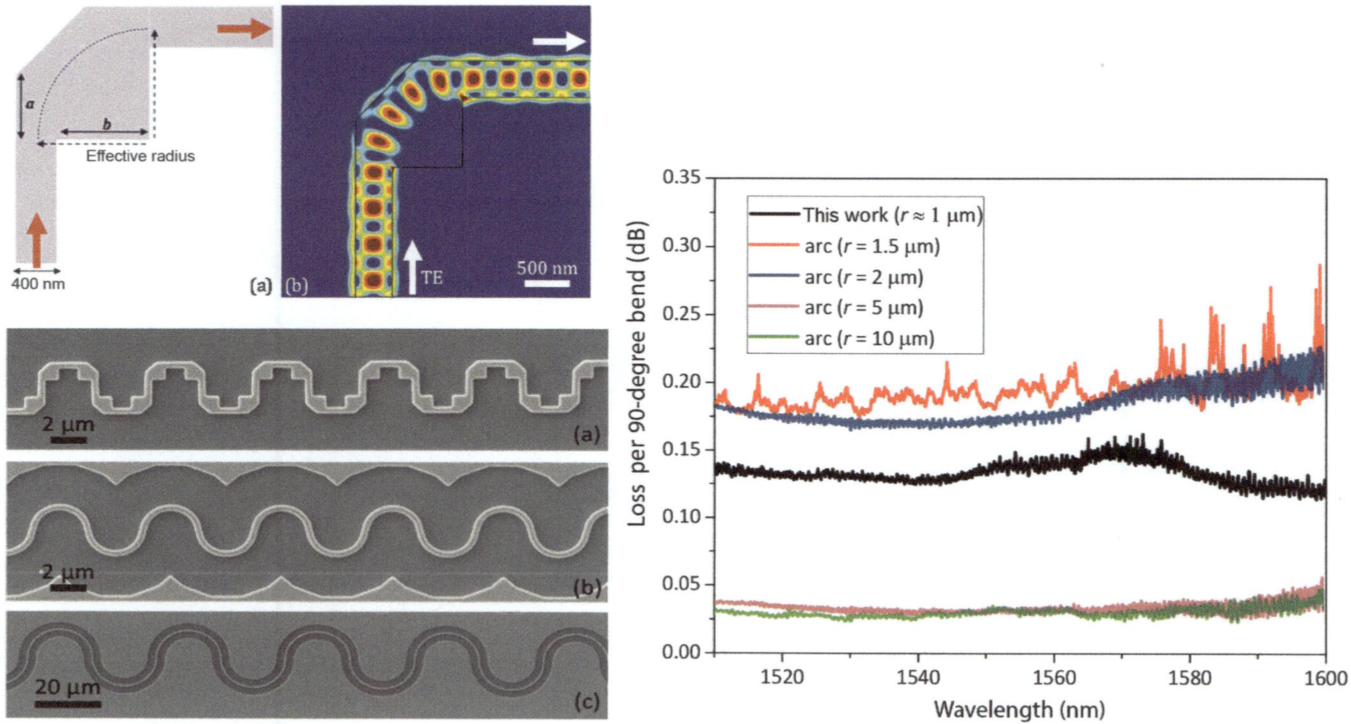

Fig. 10 Comparison of S-bends and corner mirrors. *Top left*: Structure of the corner mirror and simulation of wave propagation. *Bottom left*: SEM picture of realized devices. *Right*: experimental results (Jiao et al., 2016a).

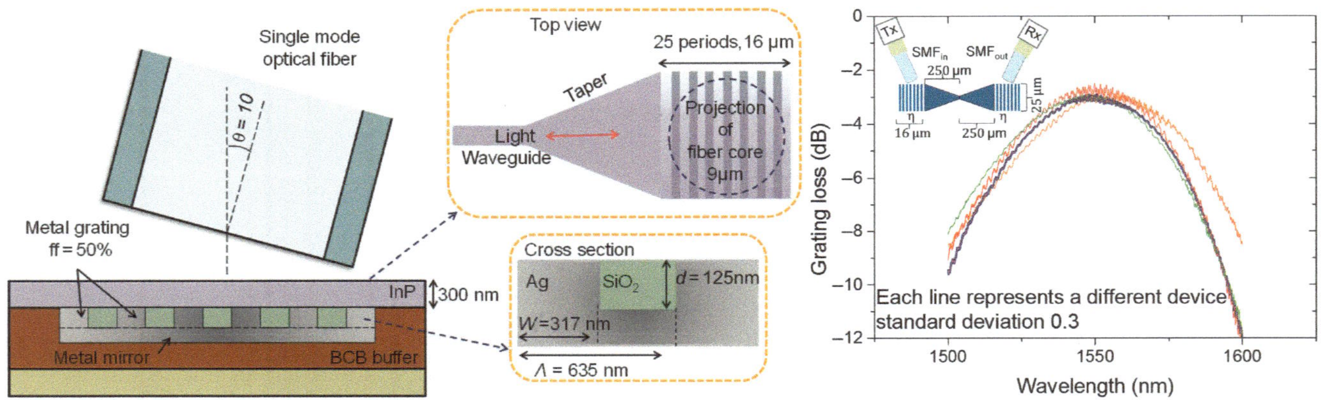

Fig. 11 Grating coupler with metal back reflector. *Left*: Structure of the device. *Right*: Transmission loss spectra (Higuera-Rodriguez et al., 2015).

are $10 \times 10\,\mu m^2$ when optimized for coupling to standard single-mode optical fiber. A 2D tapered waveguide can be used to couple the grating to the narrow membrane waveguide. A focussing grating can be obtained by curving the grating lines, so that this tapered waveguide can be avoided and a much smaller footprint is created. The optical bandwidth for these gratings is in the range of 60–100 nm, while the reflections are in the order of a few percent.

3.6 Discussion

A series of passive waveguiding devices for the InP membrane have been developed, which perform on a level comparable to other membrane platforms. Most of these are using concepts demonstrated before in low-contrast photonic integration platforms (Augustin et al., 2018; Smit et al., 2014), but the designs have to be adjusted and special care has to be taken to avoid unwanted reflections. The high-contrast membrane waveguides allow solutions which are less effective in low-contrast platforms, like the corner mirrors and the PCG. In all cases however, a large reduction in footprint is obtained for membrane devices (more than one order of magnitude in linear dimensions (Augustin et al., 2018)).

It should be mentioned that the passive devices on an InP membrane are very similar in design and performance as those on the SOI platforms (Absil et al., 2014). This is to be expected, since index contrasts are similar also. The InP membrane has the possibility of double-sided processing, providing extra flexibility, as has been used, e.g., for metal back reflector in the input/output grating. These passive devices will be combined with the active devices described in the next section to create a wide range of optical functions.

4. ACTIVE DEVICES

Active functionalities include light generation, modulation, and detection. The InP membrane offers the advantage of intrinsic gain medium integration. Active BBs are processed together with passive components in a single process flow. The absence of a substrate and the enhanced optical confinement enable opportunities in terms of efficiency and speed of operation.

4.1 Optical Amplifiers

On-chip light generation is the most important functionality for a PIC platform. For IMOS, the light needs to be generated and guided in a thin

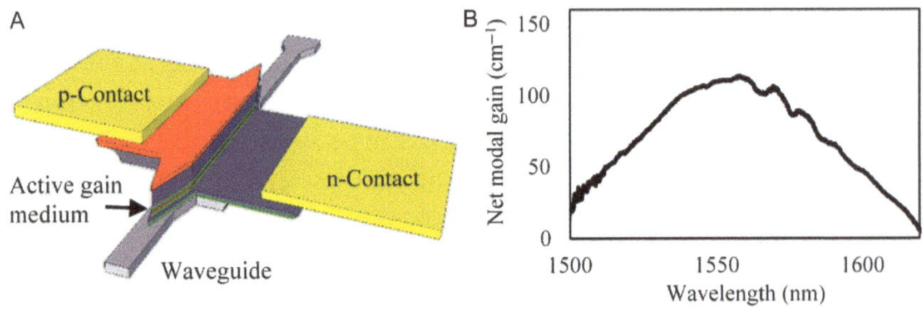

Fig. 12 (A) The twin-guide amplifier structure in InP membrane and (B) measured net modal gain (Pogoretskiy et al., 2017a).

membrane layer. A twin-guide structure is developed as shown in Fig. 12A (Pogoretskiy et al., 2017a). An active p-i-n layer stack is integrated with the 300-nm passive waveguiding layer. The active stack contains an intrinsic core with multiple quantum wells. The light generated from the active stack propagates within the active core and couples vertically to the passive guiding layer through a taper structure. This twin-guide structure is grown in a single epitaxy and is compatible with the double-sided processing technology (see Section 2). Due to the well-matched mode profile in the active and passive waveguides, a taper length of only 30 μm is needed for a 98% power transfer, as simulated with a 3D FDTD algorithm.

Gain spectra are measured in fabricated amplifier structures, using the method described in (Nazca Design™, n.d.). As can be seen in Fig. 12B, a net modal gain as high as $110\,cm^{-1}$ is achieved at the current injection density of $4\,kA/cm^2$ (Pogoretskiy et al., 2017a). The gain generated per QW in the InP membrane is comparable to that in substrate-based amplifiers (Pustakhod et al., 2018). In the next section, an example of using the InP membrane amplifier as a BB to construct a laser circuit will be presented.

4.2 Lasers

A quantum well laser in an IMOS platform has been created from the combination of high-contrast grating reflectors, surface gratings, and an optical amplifier. In order to verify the performance of the active–passive integration, a Fabry–Perot laser has been demonstrated (see Fig. 13; Pogoretskiy et al., 2017b). The laser consists of a 500-μm-long optical amplifier with taper structures at both ends to couple light vertically to and from the passive waveguides. PhC reflectors have been placed outside the taper sections,

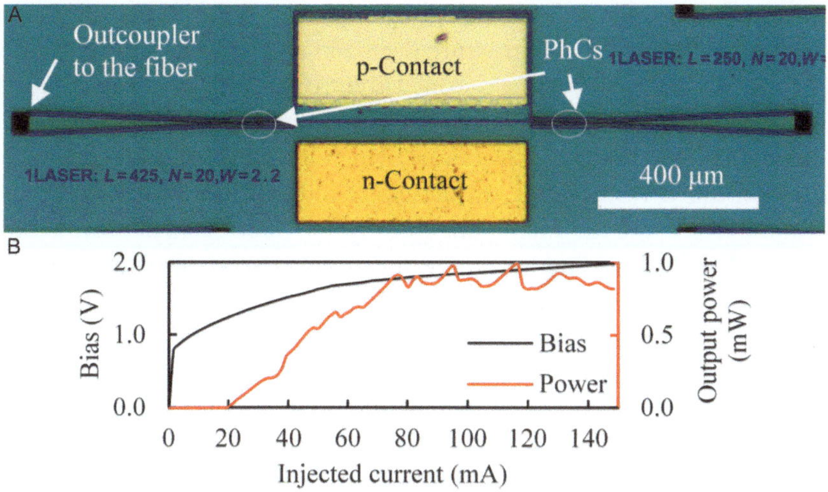

Fig. 13 Quantum well laser created using IMOS platform building blocks. (A) Microscope picture of a fabricated FP laser in InP membrane. (B) The *I–V* and *L–I* characteristics of the laser (Pogoretskiy et al., 2017a,b).

in the waveguiding layer, to form the Fabry–Perot cavity. A PhC reflector can provide high reflectivity (>95%) and wide optical bandwidth (>200 nm) with 10 PhC holes. At the end of the passive waveguide, grating couplers are placed to couple the laser light into an optical fiber. The laser diode exhibits a low series resistance (5 Ω) and threshold current (20 mA, 2 kA/cm^2). The maximum optical power in the waveguide is 1 mW. A thermal roll-over was observed at high current injection levels, indicating junction heating. This is expected to be removed with the inclusion of a thermal shunt to the silicon carrier, as shown in the platform schematic diagram of Fig. 1.

4.3 Photodetectors

Photodetectors convert optical signal into electric current. Detectors that couple with high-confinement waveguides are essential for the membrane PIC platform. A short amplifier operated at reverse bias is able to operate as a photodiode (Tol et al., 2018). The sensitivity of such an amplifier-based detector can be very good, but may be limited in terms of speed, linearity (Pello, 2014), and in the case of quantum well designs, polarization sensitivity. The relatively long transit distance for the heavy holes generated from the central intrinsic region limits the speed and can lead to charge carrier blocking.

A promising approach for high-speed and high-linearity detectors is the unitraveling carrier (UTC) structure (Ishibashi et al., 1997). The absorption region in the UTC structure is p-doped; thus the photon-generated holes do not need to transit the absorbing layers. The dynamic response is thus purely determined by the transit of electrons, which have a much higher velocity. Based on the UTC concept, an ultrafast photodetector was designed and demonstrated in the InP membrane (see Fig. 14; Shen et al., 2016a). Thanks to the high optical confinement in the membrane, the footprint of the device is only $3 \times 10\,\mu m^2$, ensuring capacitance below 10 fF. Double-sided technology was used to bring metal contacts closer and reduce the series resistance. As a result the measured 3 dB electrical bandwidth is beyond the instrument limit of 67 GHz. The predicted bandwidth using the measured capacitance and resistance is beyond 100 GHz (Shen, 2016). The detector shows good linearity in its responsivity. No optical power saturation has been observed in the measured power range up to 4 mW.

The UTC high-speed detector has been studied in a prototype communications link: an indoor optical wireless communication (OWC) system using an infrared beam. Using an integrated receiver PIC containing a UTC detector and a surface grating, 50 Gbit/s data transmission has been demonstrated with a monocolor optical free-space beam (Cao et al., 2017).

4.4 Modulators

Optical signals propagating in the waveguide can be manipulated in phase and/or intensity at high speed, using optical modulators. These are essential

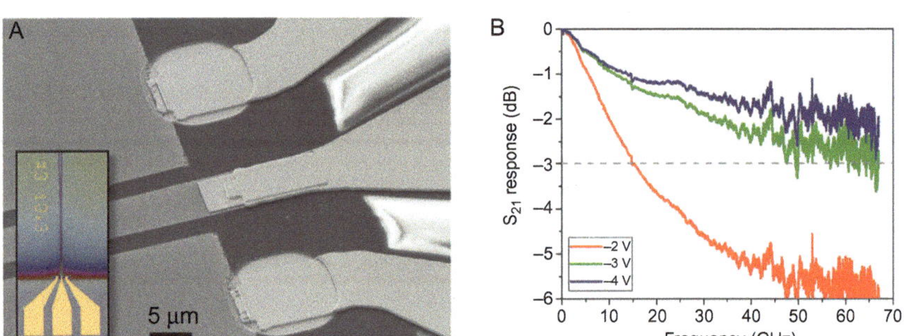

Fig. 14 (A) SEM picture and (B) measured frequency response of the UTC detector (Shen et al., 2016a).

BBs for high-speed integrated transceivers for telecommunication, data communication, and on-chip interconnects.

The phase and/or loss of the light can be modulated by rapidly changing the real and/or imaginary part of refractive index of a material that the light is interacting with. EO effects provide low optical loss and fast dynamics (Liu et al., 2015; Smit et al., 2014). EO polymers are now widely researched using silicon-based nanophotonic circuits to enable ultrahigh-speed modulation (Alloatti et al., 2014). The same concept can be adapted to the InP photonic membrane since its high optical confinement is comparable to the silicon membrane (see Section 3) to enable intimate laser integration. Slot waveguides and the structures to couple them to single-mode waveguides have been successfully developed (Mejia, 2018; Millan Mejia et al., 2016). The cross-sectional structure is shown in Fig. 15, indicating the slots in the two branches of a Mach–Zehnder interferometer and a GSG-type electrode structure to apply high-frequency electrical fields to the polymer. The application of EO polymer to the InP membrane slot modulator is being investigated at the time of writing this chapter (Mejia, 2018).

Modulation of light intensity typically exploits the QCSE. The wavelength of the light and the band edge of the QWs have to be detuned by a critical amount, resulting in a narrow-optical bandwidth of only few nanometers for EAM. An EAM with wide-optical bandwidth can be obtained using the band-filling effect in a highly n-doped InGaAs absorption layer (Engelen et al., 2018). A schematic structure of the modulator is shown in Fig. 16, indicating an S-shaped diode structure with p- and n-contacts on opposite sides of the waveguide core. The light absorption

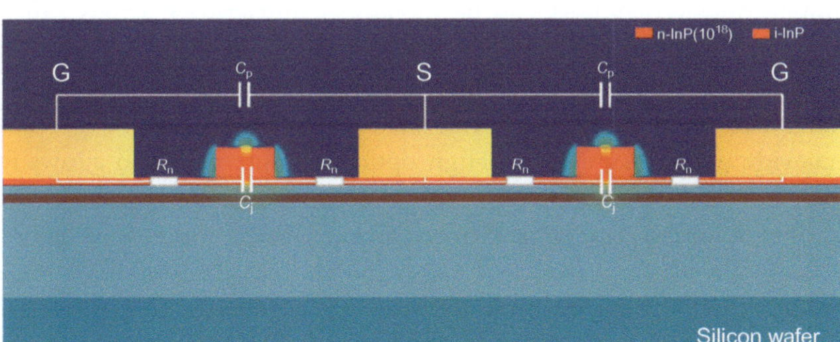

Fig. 15 Cross-section of the slot modulator with EO polymer (Mejia, 2018).

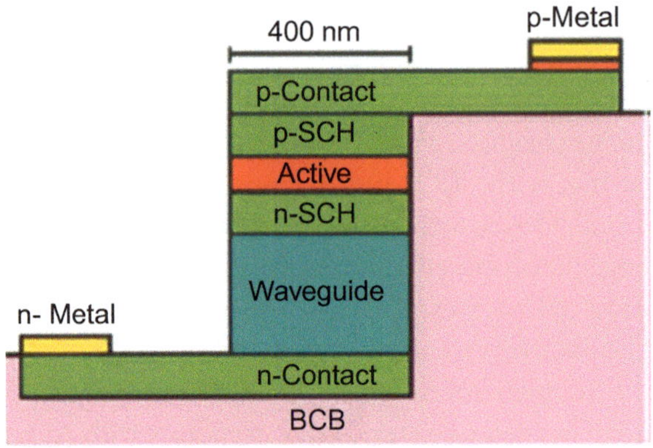

Fig. 16 Cross-section of the band-filling effect EAM (Engelen et al., 2018).

strength can be rapidly modulated by the changing electron concentration under electric bias. A design trade-off lies between the extinction ratio and the operation speed of the device. With a contact area of $0.4 \times 80\,\mu m^2$, the device is predicted to have a static extinction ratio of 7.2 dB with a 4.4 dB on-state insertion loss. It has been predicted to operate with 35 GHz RC-limited electrical bandwidth and 160 nm optical bandwidth. This is the result of the wavelength independent nature of the modulating mechanisms, which is switching on and off the material absorption of InGaAs.

4.5 Discussion

The highly confined InP photonic membrane enables a series of novel active devices. High-performance optical amplifiers integrated with nanophotonic waveguides can be directly manufactured. Lasers can be constructed based on the amplifiers together with broadband PhC mirrors. Innovations in the layer structures to tailor carrier dynamics result in an ultrafast photodetector beyond 67 GHz and an EAM design with ultra-wide optical bandwidth. High optical confinement also allows for strong light matter interaction, which is beneficial for applying novel photonic materials (e.g., an EO polymer) on nanophotonic waveguides.

From the platform point of view, integrating different layer stacks (amplifier, UTC detector, and band-filling EAM) can be a challenge. The butt-joint regrowth technology is mature and widely used in generic InP foundry processes (Augustin et al., 2018). This can be a promising technology to extend the InP membrane platform with multiple types of functional layer stacks.

5. FUTURE DIRECTIONS

The membrane platform, with its high-speed and high-density characteristics, offers the opportunity for circuit designers to combine high-performance BBs with a high degree of flexibility. One important class of nanophotonic circuit is the millimeter wave beam steering element (Pello, 2014), highly promising for next-generation wireless networks. The fully integrated heterodyne generation of millimeter waves on an InP membrane chip will require highly stable lasers, ultrafast UTC PDs, and on-chip antennas. The possibility of dense integration facilitates large-scale phased arrays. OWC offers the promise of secure, high-bandwidth in-door communications. Surface-grating couplers can act as compact, efficient optical antennas. Further integration with lasers and detectors will enable OWC transmitters and receivers on a single chip. A first demonstration of a high-speed OWC receiver membrane circuit is given in Cao et al. (2017). The IMOS platform can additionally be a promising solution for on-chip sensing (Jiao et al., 2016b, 2017). The strong field of the modes at the interfaces of the high confinement waveguides makes them suitable for on-chip optical sensing, by detecting adsorbed molecules.

The InP membrane platform, IMOS, is becoming available to support all these, and more, applications. An important choice is made by opening up the technology for designers. These can apply for a cell on a wafer, on which they can have their intended photonic circuits realized in an MPW (multiproject wafer run). This is supported by making available design BBs, representing the actual devices on the platform. These BBs are put together in a PDK library. The next section describes this.

Comparisons with other platforms (SOI (Absil et al., 2014), generic integration (Smit et al., 2014)) show that improved performances are possible. Moreover, the potential that the membrane brings with respect to cointegration of photonics and electronics has not been explored yet. Also, applying specific features of this platform, such as the high-power density or the strong interaction with interfaces, provides future research directions. These issues will be addressed in the subsequent sections.

5.1 Process Design Kit

The PDK contains technology- and/or foundry-specific information (Smit et al., 2014). This provides a layer of abstraction for mask layout and includes information on BBs, design rules of PICs, and templates for optical and electrical connectivity. An overview of some experimentally demonstrated BBs is given in Table 1.

Table 1 List of Experimentally Demonstrated BBs Available in the PDK

Building Block	Parameter	Performance
Waveguide	Propagation loss	3–5 dB/cm
Bent waveguide	Insertion loss	0.03 dB/90 degrees at 5 μm radius
90-degree bend with sharp mirror	Effective radius	0.96 μm
	Insertion loss	0.13 dB/90 degrees
AWG	Insertion loss	10 dB
	Cross talk	10 dB
	Tunability	Thermo-optic
PCG	Insertion loss	4 dB
	Cross-talk	25 dB
	Tunability	Thermo-optic
Ring resonator	Q factor	Up to 40,000
	Tunability	Thermo-optic
1×2 MMI coupler	Insertion loss	0.6 dB
Linear grating coupler	Insertion loss	5.5 dB to SMF (TE)
	Optical 3 dB bandwidth	100 nm
Focusing grating coupler	Insertion loss	4.5 dB to SMF (TE and TM)
	Optical 3 dB bandwidth	100 nm
Thermo-optic phase shifter	Length for 2π phase shift	200 μm (metal width of 2 μm and ΔT of 50°C)
SOA	Net model gain (TE)	110 cm^{-1} @ 4 kA/cm^2

Current development of PDK library for the InP membrane platform focuses on the mask library. The above-listed membrane devices have been parameterized as a BB. To build a circuit, one only needs to call the BB function in a layout editor and place it at the designed location. BBs can be connected by waveguides, bends, and other interconnections. The membrane PDK can be developed in commercially available software frameworks such as Nazca Design (Nazca Design™, n.d.) and Synopsys OptoDesigner (Synopsys (Phoenix Software), n.d.). Mask layouts of a wide range of active and passive BBs have been made available, including

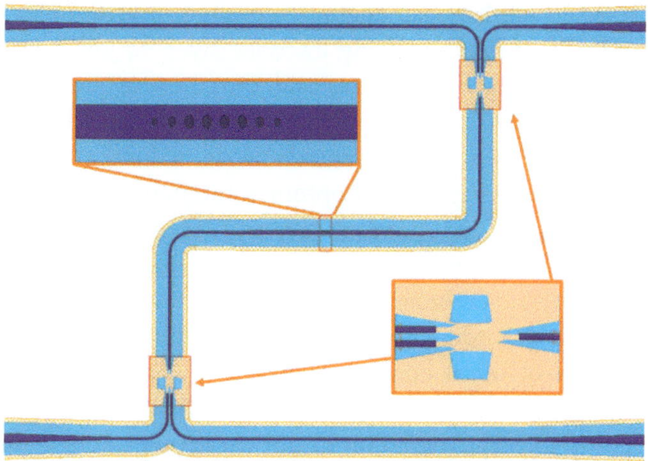

Fig. 17 An example mask layout of a membrane circuit for reflectivity measurement (Pello, 2014). Building blocks are indicated by the *red square* and are interconnected by waveguide circuits.

the twin-guide amplifiers and lasers. An example mask layout of a membrane circuit for reflectivity measurement (Pello, 2014) is shown in Fig. 17.

With the mask library, designers can create complex photonic circuits without needing to know the detailed layer structure and process technology. This can facilitate accessibility to new designers and new applications for the InP membrane platform.

5.2 Technological Developments

To enable the growth of both the scale and the functionality on the InP membrane platform a number of advanced techniques will be applied. One example of this is an improved epitaxial growth. The present results have been obtained with MOCVD, growing first the n-doped layer and finishing with the p-doped layers. This is necessary to limit diffusion of Zn, the p-dopant, into the active materials. There are however advantages in growing the diode in an inverted way, e.g., from a thermal perspective, since the dominant ohmic heating from the p-doped layers can be more easily removed from the top of the bonded membrane (see Fig. 1). Such an inverted growth can be realized with a different epitaxy (in particular with MBE), by using other dopants (like beryllium), or by blocking the Zn-diffusion. This latter option is quite interesting, as has been demonstrated in Daunt et al. (2012).

On a device level, bandgap engineering offers improvement of performance. Band smoothing, by adding interface layers to junctions, is shown in simulations to reduce resistance, and thus heating, in active devices (van Engelen et al., 2017a). Furthermore, since in the membrane geometry metal contacts are close to the waveguide mode, reduction of optical losses in these contacts allows greatly to optimize the designs. As mentioned in Section 2, n-type contacts using Ag/Ge-based contacts provide comparable electrical performance as more traditional Au/Ni-based contacts (with specific contact resistance in the range of $10^{-7}\,\Omega\,\mathrm{cm}^2$), but show an optical loss that is more than an order of magnitude lower (0.2 dB/μm). This allows closer placement of metal contacts and thus lower series resistances and higher frequency operation. Finally, also the heatsinking of the active devices needs to be improved. This can be done by making a metal bridge with the substrate (van Engelen et al., 2017b; Zhang et al., 2015), which can improve the power-handling capacity of the membrane devices by at least a factor of 3.

On the circuit level the platform will be expanding its possibilities. An alternative to the use of diffractive gratings (see Section 3) is to use an inverted taper, like those demonstrated in SOI (Pu et al., 2010), to allow butt-joint fiber coupling. In that case packaging approaches that are already developed for generic integration platforms can be applied (Smit et al., 2014). Furthermore, the platform functionality will include polarization handling. While one basic function for this, a polarization converter, has been demonstrated, another essential requirement is the dynamic control of the phase difference between the TE- and TM-like modes, which will be developed as well.

5.3 Integration Platform

The use of a twin-guide approach, as described in Section 2, has a clear advantage in needing only a single epitaxial growth. There are also drawbacks, however, like the use of tapers for the active–passive coupling and limitations in layer stack design. An alternative option is that the platform can be based on regrowth, as is used in generic integration platforms (Smit et al., 2014). This is challenging in a high-index contrast membrane. It is required to investigate this route however, if miniaturization and device optimization must be brought further. Especially this will allow smallest footprints optimization of lasers, detectors, modulators, and passive devices.

Another point is that the platform has some advantageous properties, and some of them shared with the SOI platforms, opening opportunities for new functions. The strong evanescent fields of the modes allow interaction with materials on the waveguide interfaces. An example is the use of magneto-optic materials (Koopmans et al., 2018), suggesting integration of non-reciprocal devices and optical memories. Another example is the use of 2D materials, which can be studied and exploited through interactions with the strong evanescent fields of the membrane waveguide modes. Furthermore, the high confinement waveguides result in high-power densities, which might be exploited for nonlinear optical functionality.

6. CONCLUSIONS

Recent technology developments in the InP photonic membrane platform have enabled the combination of high-performance amplifiers and detectors with enhanced optical confinement and reduced critical dimension passive devices on a wafer-scale platform.

A series of passive waveguiding devices for the InP membrane have been developed, which perform at the state-of-the-art level. High-contrast membrane waveguides allow solutions which would otherwise be less effective in low-contrast platforms. Examples include the corner mirrors and the PCG. In all cases however, a large reduction in footprint is obtained for membrane devices.

Lasers can be constructed based on the amplifiers together with broadband PhC mirrors. Innovations in the layer structures in an ultrafast photodetector beyond 67 GHz and an EAM design with ultra-wide optical bandwidth. High optical confinement also allows for strong light–matter interaction, which is beneficial for applying novel photonic materials (e.g., EO polymer) on nanophotonic waveguides.

The methodology followed for platform integration has been a critical enabler for increased InP circuit developments [JePPIX.eu] and we anticipate a similar trajectory for IMOS technologies. The first PDK is currently under development and is expected to enable a broader range of designers to develop circuits from the powerful set of IMOS BBs.

REFERENCES

Abbasi, A., Spatharakis, C., Kanakis, G., Andr, N.S., Louchet, H., Katumba, A., Verbist, J., Avramopoulos, H., Bienstman, P., Yin, X., Bauwelinck, J., Roelkens, G., Morthier, G., 2016. High speed direct modulation of a heterogeneously integrated InP/SOI DFB laser. J. Lightwave Technol. 34, 1683–1687.

Absil, P., Verheyen, P., Pantouvaki, M., Heyn, P.D., Chen, H., Lepage, G., Coster, J.D., Rakowski, M., Dumon, P., Khanna, A., Roelkens, G., Baets, R., Campenhout, J.V., 2014. In: Review of a 200mm SOI silicon photonics platform performance for low-power optical integrated circuits development. 4th International Symposium on Photonics and Electronics Convergence—Tokyo, Japan.

Alloatti, L., Palmer, R., Diebold, S., Pahl, K.P., Chen, B., Dinu, R., Fournier, M., Fedeli, J.-M., Zwick, T., Freude, W., Koos, C., Leuthold, J., 2014. 100GHz silicon-organic hybrid modulator. Light Sci. Appl. 3, e173.

Arakawa, Y., Nakamura, T., Urino, Y., Fujita, T., 2013. Silicon photonics for next generation system integration platform. IEEE Commun. Mag. 51, 72–77.

Augustin, L.M., Santos, R., den Haan, E., Kleijn, S., Thijs, P.J.A., Latkowski, S., Zhao, D., Yao, W., Bolk, J., Ambrosius, H., Mingaleev, S., Richter, A., Bakker, A., Korthorst, T., 2018. InP-based generic foundry platform for photonic integrated circuits. IEEE J. Sel. Top. Quantum Electron. 24, 1–10.

Barbarin, Y., Bente, E.A.J.M., Marquet, C., Leclere, E.J.S., Binsma, J.J.M., Smit, M.K., 2005. Measurement of reflectivity of butt-joint active-passive interfaces in integrated extended cavity lasers. IEEE Photon. Technol. Lett. 17, 2265–2267.

Burdeaux, D., Townsend, P., Carr, J., Garrou, P., 1990. Benzocyclobutene (BCB) dielectrics for the fabrication of high density, thin film multichip modules. J. Electron. Mater. 19, 1357–1366.

Cao, Z., Jiao, Y., Shen, L., Zhao, X., Stabile, R., van der Tol, J.J.G.M., Koonen, T., 2017. Ultra-high throughput indoor infrared wireless communication system enabled by a cascaded aperture optical receiver fabricated on InP membrane. J. Lightwave Technol. 36, 57–67.

Corbett, B., Loi, R., Zhou, W., Liu, D., Ma, Z., 2017. Transfer print techniques for heterogeneous integration of photonic components. Prog. Quantum Electron. 52, 1–17.

Daunt, C.L.M., Cleary, C.S., Manning, R.J., Thomas, K., Young, R.J., Pelucchi, E., Corbett, B., Peters, F.H., 2012. Sub 10 ps carrier response times in electroabsorption modulators using quantum well offsetting. IEEE J. Quantum Electron. 48, 1467–1475.

Davenport, M.L., Skendzic, S., Volet, N., Hulme, J.C., Heck, M.J.R., Bowers, J.E., 2016. Heterogeneous silicon/III-V semiconductor optical amplifiers. IEEE J. Sel. Top. Quantum Electron. 22, 3100111.

Dolores-Calzadilla, V., Heiss, D., Smit, M., 2014. Highly efficient metal grating coupler for membrane-based integrated photonics. Opt. Lett. 39, 2786–2789.

Dolores-Calzadilla, V., Romeira, B., Pagliano, F., Birindelli, S., Higuera-Rodriguez, A., van Veldhoven, P.J., Smit, M.K., Fiore, A., Heiss, D., 2017. Waveguide-coupled nanopillar metal-cavity light-emitting diodes on silicon. Nat. Commun. 8, 14323.

Duan, G.H., Jany, C., Liepvre, A.L., Accard, A., Lamponi, M., Make, D., Kaspar, P., Levaufre, G., Girard, N., Lelarge, F., Fedeli, J.M., Descos, A., Bakir, B.B., Messaoudene, S., Bordel, D., Menezo, S., Valicourt, G.d., Keyvaninia, S., Roelkens, G., Thourhout, D.V., Thomson, D.J., Gardes, F.Y., Reed, G.T., 2014. Hybrid III–V on silicon lasers for photonic integrated circuits on silicon. IEEE J. Sel. Top. Quantum Electron. 20, 158–170.

Engelen, J.P.v., Shen, L., Roelkens, G., Jiao, Y., Smit, M.K., van der Tol, J.J.G.M., 2018. A novel broadband electro-absorption modulator based on bandfilling in n-InGaAs: design and simulations. IEEE J. Sel. Top. Quantum Electron. 24, 1–8.

Fujii, T., Takeda, K., Diamantopoulos, N.P., Kanno, E., Hasebe, K., Nishi, H., Nakao, R., Kakitsuka, T., Matsuo, S., 2018. Heterogeneously integrated membrane lasers on Si substrate for low operating energy optical links. IEEE J. Sel. Top. Quantum Electron. 24, 1–8.

Higuera-Rodriguez, A., Dolores-Calzadilla, V., Jiao, Y., Geluk, E.J., Heiss, D., Smit, M.K., 2015. Realization of efficient metal grating couplers for membrane-based integrated photonics. Opt. Lett. 40, 2755–2757.

Inoue, D., Hiratani, T., Fukuda, K., Tomiyasu, T., Amemiya, T., Nishiyama, N., Arai, S., 2016. Low-bias current 10 Gbit/s direct modulation of GaInAsP/InP membrane DFB laser on silicon. Opt. Express 24, 18571–18579.

Ishibashi, T., Shimizu, N., Kodama, S., Ito, H., Nagatsuma, T., Furuta, T., 1997. Uni-traveling-carrier photodiodes. In: Nuss, M., Bowers, J. (Eds.), Ultrafast Electronics and Optoelectronics. OSA Trends in Optics and Photonics Series, 13, Optical Society of America, p. UC3.

Ishii, T., Shigehara, K., 2009. Fullerene nanocomposite resists for nanolithography. In: Merhari, L. (Ed.), Hybrid Nanocomposites for Nanotechnology. Springer.

Jain, S.R., Sysak, M.N., Kurczveil, G., Bowers, J.E., 2011. Integrated hybrid silicon DFB laser-EAM array using quantum well intermixing. Opt. Express 19, 13692–13699.

Jiao, Y., Pello, J., Mejia, A.M., Shen, L., Smalbrugge, B., Geluk, E.J., Smit, M., van der Tol, J., 2014. Fullerene-assisted electron-beam lithography for pattern improvement and loss reduction in InP membrane waveguide devices. Opt. Lett. 39, 1645–1648.

Jiao, Y., de Vries, T., Unger, R.-S., Shen, L., Ambrosius, H., Radu, C., Arens, M., Smit, M., van der Tol, J., 2015. Vertical and smooth single-step reactive ion etching process for InP membrane waveguides. J. Electrochem. Soc. 162, E90–E95.

Jiao, Y., Liu, J., Mejia, A.M., Shen, L., Tol, J.v.d., 2016a. Ultra-sharp and highly tolerant waveguide bends for InP photonic membrane circuits. IEEE Photon. Technol. Lett. 28, 1637–1640.

Jiao, Y., Shen, L., Cao, Z., Latkowski, S., Bente, E.A.J.M., van der Tol, J.J.G.M., Smit, M.K., Williams, K.A., 2016b. In: III-V photonic integrated circuits for beyond-telecom applications. 25th Wireless and Optical Communication Conference (WOCC 2016).

Jiao, Y., van der Tol, J.J.G.M., Shen, L., Mejia, A.M., Ambrosius, H.P.M.M., Smit, M.K., Williams, K.A., 2017. In: High confinement InP nanophotonic circuits for optical sensing opportunities. 38th Progress In Electromagnetics Research Symposium (PIERS 2017).

Karouta, F., 2014. A practical approach to reactive ion etching. J. Phys. D Appl. Phys. 47, 233501.

Kish, F., Lal, V., Evans, P., Corzine, S.W., Ziari, M., Butrie, T., Reffle, M., Tsai, H.S., Dentai, A., Pleumeekers, J., Missey, M., Fisher, M., Murthy, S., Salvatore, R., Samra, P., Demars, S., Kim, N., James, A., Hosseini, A., Studenkov, P., Lauermann, M., Going, R., Lu, M., Zhang, J., Tang, J., Bostak, J., Vallaitis, T., Kuntz, M., Pavinski, D., Karanicolas, A., Behnia, B., Engel, D., Khayam, O., Modi, N., Chitgarha, M.R., Mertz, P., Ko, W., Maher, R., Osenbach, J., Rahn, J.T., Sun, H., Wu, K.T., Mitchell, M., Welch, D., 2018. System-on-chip photonic integrated circuits. IEEE J. Sel. Top. Quantum Electron. 24, 1–20.

Koopmans, B., van Hees, Y.L.W., Lalieu, M.L.M., Demirer, F.E., van der Tol, J.J.G.M., Lavrijsen, R., 2018. In: Towards integrated magnetophotonics using claddings with perpendicular magnetic anisotropy. 20th European Conference on Integrated Optics (ECIO 2018).

Kurczveil, G., Pintus, P., Heck, M.J.R., Peters, J.D., Bowers, J.E., 2013. Characterization of insertion loss and back reflection in passive hybrid silicon tapers. IEEE Photon. J. 5, 6600410.

Liu, K., Ye, C.R., Khan, S., Sorger, V.J., 2015. Review and perspective on ultrafast wavelength-size electro-optic modulators. Laser Photon. Rev. 9, 172–194.

Matsuo, S., Fujii, T., Hasebe, K., Takeda, K., Sato, T., Kakitsuka, T., 2015. Directly modulated DFB laser on SiO_2/Si substrate for datacenter networks. J. Lightwave Technol. 33, 1217–1222.

Mejia, A.J.M., 2018. Development of IMOS Technology for a High Bandwidth Modulator. PhD Thesis. Department of Electrical Engineering, Technische Universiteit Eindhoven, Eindhoven, The Netherlands.

Millan Mejia, A.J., Jiao, Y., van der Tol, J.J.G.M., Smit, M.K., 2016. In: Fabrication technology of a slot waveguide modulator in InP membranes on silicon (IMOS). 18th European Conference on Integrated Optics (ECIO 2016).

Millan-Mejia, A.J., Jiao, Y., van der Tol, J.J.G.M., Smit, M.K., 2016. In: Design of a deep etched focusing subwavelength grating with a metallic reactor for membrane based photonic integration platforms. Proceedings of the 21st Annual Symposium of the IEEE Photonics Society Benelux Chapter.
Nazca Design^{TM}, n.d. http://nazca-design.org.
Niklaus, F., Stemme, G., Lu, J.-Q., Gutmann, R.J., 2006. Adhesive wafer bonding. J. Appl. Phys. 99, 031101.
Pello, J., 2014. Building up a Membrane Photonics Platform in Indium Phosphide. PhD Thesis. Department of Electrical Engineering, Eindhoven University of Technology, Eindhoven, The Netherlands.
Pello, J., van der Tol, J., Keyvaninia, S., van Veldhoven, R., Ambrosius, H., Roelkens, G., Smit, M., 2012. High-efficiency ultrasmall polarization converter in InP membrane. Opt. Lett. 37, 3711–3713.
Pello, J., Muneeb, M., Keyvaninia, S., van der Tol, J.J.G.M., Roelkens, G., Smit, M.K., 2013. Planar concave grating demultiplexers on an InP-membrane-on-silicon photonic platform. IEEE Photon. Technol. Lett. 25, 1969–1972.
Pogoretskiy, V., van Engelen, J., van der Tol, J., Higuera-Rodriguez, A., Smit, M., Jiao, Y., 2017a. An integrated SOA-building block for an InP-membrane platform. In: Advanced Photonics 2017 (IPR, NOMA, Sensors, Networks, SPPCom, PS). p. JW4A.1. Optical Society of America, New Orleans, Louisiana.
Pogoretskiy, V., Tol, J.v.d., Higuera-Rodriguez, A., Smit, M., Jiao, Y., 2017b. In: Integrated photonic crystal DBR laser in an InP membrane platform. 39th Progress in Electromagnetic Research Symposium (PIERS 2017), Singapore.
Pu, M., Liu, L., Ou, H., Yvind, K., Hvam, J.M., 2010. Ultra-low-loss inverted taper coupler for silicon-on-insulator ridge waveguide. Opt. Commun. 283, 3678–3682.
Pustakhod, D., Williams, K., Leijtens, X., 2018. Fast and robust method for measuring semiconductor optical amplifier gain. IEEE J. Sel. Top. Quantum Electron. 24, 1–9.
Rouvalis, E., Chtioui, M., van Dijk, F., Lelarge, F., Fice, M.J., Renaud, C.C., Carpintero, G., Seeds, A.J., 2012. 170 GHz uni-traveling carrier photodiodes for InP-based photonic integrated circuits. Opt. Express 20, 20090–20095.
Shen, L., 2016. Ultrafast Photodetector on the InP-Membrane-On-Silicon Platform. PhD Thesis. Department of Electrical Engineering, Technische Universiteit Eindhoven, Eindhoven, The Netherlands.
Shen, L., Dolores-Calzadilla, V., Wullems, C.W.H.A., Jiao, Y., Millan-Mejia, A., Higuera-Rodriguez, A., Heiss, D., van der Tol, J.J.G.M., Ambrosius, H.P.M.M., Roelkens, G., Smit, M.K., 2015. Low-optical-loss, low-resistance Ag/Ge based ohmic contacts to n-type InP for membrane based waveguide devices. Opt. Mater. Express 5, 393–398.
Shen, L., Jiao, Y., Yao, W., Cao, Z., Engelen, J.P.v., Tol, J.J.G.M.v.d., Roelkens, G., Smit, M.K., 2016a. High-bandwidth uni-traveling carrier waveguide photodetector on an InP-membrane-on-silicon platform. Opt. Express 24, 8290–8301.
Shen, L., Jiao, Y., Rodriguez, A.H., Mejia, A.J.M., Roelkens, G.C., Smit, M.K., van der Tol, J.J.G.M., 2016b. In: Double-sided processing for membrane-based photonic integration. 18th European Conference on Integrated Optics (ECIO 2016), Warsaw, Poland.
Shen, L., van Veldhoven, P.J., Jiao, Y., Dolores-Calzadilla, V., van der Tol, J.J.G.M., Roelkens, G., Smit, M.K., 2016c. Ohmic contacts with ultra-low optical loss on heavily doped n-type InGaAs and InGaAsP for InP-based photonic membranes. IEEE Photon. J. 8, 1–10.
Smit, M., Leijtens, X., Ambrosius, H., Bente, E., Tol, J.v.d., Smalbrugge, B., Vries, T.d., Geluk, E.-J., Bolk, J., Veldhoven, R.v., Augustin, L., Thijs, P., D'Agostino, D., Rabbani, H., Lawniczuk, K., Stopinski, S., Tahvili, S., Corradi, A., Kleijn, E., Dzibrou, D., Felicetti, M., Bitincka, E., Moskalenko, V., Zhao, J., Santos, R., Gilardi, G., Yao, W., Williams, K., Stabile, P., Kuindersma, P., Pello, J., Bhat, S.,

Jiao, Y., Heiss, D., Roelkens, G., Wale, M., Firth, P., Soares, F., Grote, N., Schell, M., Debregeas, H., Achouche, M., Gentner, J.-L., Bakker, A., Korthorst, T., Gallagher, D., Dabbs, A., Melloni, A., Morichetti, F., Melati, D., Wonfor, A., Penty, R., Broeke, R., Musk, B., Robbins, D., 2014. An introduction to InP-based generic integration technology. Semicond. Sci. Technol. 29, 083001.

Synopsys (Phoenix Software). n.d. Available: https://www.phoenixbv.com/.

Tol, J.J.G.M.v.d., Jiao, Y., Shen, L., Millan-Mejia, A., Pogoretskii, V., Engelen, J.P.v., Smit, M.K., 2018. Indium phosphide integrated photonics in membranes. IEEE J. Sel. Top. Quantum Electron. 24, 6100809.

van der Tol, J., Pello, J., Bhat, S., Jiao, Y., Heiss, D., Roelkens, G., Ambrosius, H., Smit, M., 2014. In: Photonic integration in indium-phosphide membranes on silicon (IMOS). Proceedings of SPIE 8988, Integrated Optics: Devices, Materials, and Technologies XVIII. 89880M–89880M-17.

van Engelen, J.P., et al., 2017a. In: Improving thermal performance of a UTC photodetector in the IMOS platform. 19th European Conference on Integrated Optics (ECIO 2017).

van Engelen, J.P., Pogoretskii, V., Smit, M.K., van der Tol, J.J.G.M., Jiao, Y., 2017b. In: Towards a fully integrated indium-phosphide membrane on silicon photonics platform. Proceedings of the 22nd Annual Symposium of the IEEE Photonics Society Benelux Chapter, Delft, The Netherlands.

Wang, Z., Tian, B., Pantouvaki, M., Guo, W., Absil, P., Van Campenhout, J., Merckling, C., Van Thourhout, D., 2015. Room-temperature InP distributed feedback laser array directly grown on silicon. Nat. Photonics 9, 837.

Xie, X., Zhou, Q., Li, K., Shen, Y., Li, Q., Yang, Z., Beling, A., Campbell, J.C., 2014. Improved power conversion efficiency in high-performance photodiodes by flip-chip bonding on diamond. Optica 1, 429–435.

Zhang, C., Liang, D., Kurczveil, G., Bowers, J.E., Beausoleil, R.G., 2015. Thermal management of hybrid silicon ring lasers for high temperature operation. IEEE J. Sel. Top. Quantum Electron. 21, 385–391.

Zhang, J., Haq, B., O'Callaghan, J., Gocalinska, A., Pelucchi, E., Trindade, A.J., Corbett, B., Morthier, G., Roelkens, G., 2018. Transfer-printing-based integration of a III-V-on-silicon distributed feedback laser. Opt. Express 26, 8821–8830.

Zhu, Y., Wang, J., Xie, W., Tian, B., Li, Y., Brainis, E., Jiao, Y., Van Thourhout, D., 2017. Ultra-compact silicon nitride grating coupler for microscopy systems. Opt. Express 25, 33297–33304.

INDEX

Note: Page numbers followed by "*f*" indicate figures, and "*t*" indicate tables.

T

U

V

W

X

FSC
www.fsc.org
MIX
Papier | Fördert gute Waldnutzung
FSC® C083411

Kolibri 360 GmbH
Ferdinand-Jühlke-Straße 7
99095 Erfurt, Deutschland
produktsicherheit@kolibri360.de